THE ECOLOGY OF
MIGRANT
BIRDS

W9-BHM-732

THE ECOLOGY OF MIGRANT BIRDS

A Neotropical Perspective

JOHN H. RAPPOLE

SMITHSONIAN INSTITUTION PRESS

WASHINGTON AND LONDON

QL685.7 .R36 1995

Rappole, John H.

The ecology of migrant
  birds : a Neotropical

© 1995 by the Smithsonian Institution

Editor: Rosemary Sheffield
Production editor: Deborah L. Sanders
Designer: Kathleen Sims

Library of Congress Cataloging-in-Publication Data
Rappole, John H.
The ecology of migrant birds : a neotropical perspective / John H. Rappole.
        p.   cm.
   Includes bibliographical references (p.   ) and index.
   ISBN 1-56098-514-3 (cloth).
   1. Birds—Ecology—Latin America.   2. Birds—Wintering—Latin America.   3. Birds—
Migration—America.   4. Birds—America.   I. Title.
QL685.7.R36   1995
598.252'5—dc20   95-10214

British Library Cataloguing-in-Publication Data available
A paperback reissue (ISBN 1-56098-513-5) of the original cloth edition.
Printed in the United States of America
00 99 98 97   5 4 3 2 1

 ∞ The paper used in this publication meets the minimum requirements of the American
National Standard for Permanence of Paper for Printed Library Materials Z39.48-1984.

To my beloved wife,
Bonnie Elizabeth Carlson Rappole

CONTENTS

FIGURES AND TABLES

FIGURES

TABLES

PREFACE

This book represents a considerable expansion of a project that was first proposed by Gene Morton of the National Zoological Park, Smithsonian Institution, and Tom Lovejoy, then of the World Wildlife Fund. The initial work, a summary of available information on migratory birds in the Neotropics, was published by the U.S. Fish and Wildlife Service in 1983 as *Nearctic Avian Migrants in the Neotropics,* with Gene Morton, Tom Lovejoy, and Jim Ruos, then of the U.S. Fish and Wildlife Service, as my coauthors. A revised edition was published in Spanish by the Conservation and Research Center of the National Zoological Park in 1993 with funding assistance from the Smithsonian's Office of the Assistant Secretary for Science. Work on a second edition in English was begun in 1990, but it quickly became apparent that the revision was going to be so extensive that a new book was warranted.

This new book contains some updated tables and figures from the 1983 and 1993 books, as well as a number of new tables and figures. The principal change has been in the text, which includes several new topics and expanded coverage of topics treated in earlier editions. The entire field of migrant study has been transformed in the past decade by a quantum increase in interest in migratory birds among government agencies, conservation organizations, and the scientific community. I have incorporated as many as possible of the findings generated by this newly awakened interest and placed them in perspective. I hope that readers will find these additions to be useful and informative.

The publishers and I agreed that this new book, with its extensive text, tables, and figures, would have been too cumbersome if it had included updates of all of the appendixes and maps and the bibliography from the earlier versions. Therefore several of those sections have been omitted. I hope to publish them elsewhere, but in the interim they are available in the Spanish edition or from me on computer disk for a modest fee. (For more information, write me at Conservation and Research Center, 1500 Remount Road, Front Royal VA 22630.)

Though I am responsible for the prose and opinions expressed in this work,

I have used the plural pronoun "we" often throughout the text. Its use reflects the fact that most of the work I have done on migrants in my career has been collaborative, with extensive input from colleagues and students. In particular, Mario Ramos, Dwain Warner, Gene Morton, and Kevin Winker have had a great deal to do with much of the fieldwork as well as a number of the ideas expressed herein on the biology of migratory species.

ACKNOWLEDGMENTS

The staffs of the University of Georgia Science Library and the Smithsonian Institution's National Zoological Park Library, especially Nell Evans and Kay Kenyon, were very helpful in expediting location of pertinent literature by myself and my assistants. Candy Yelton, Linda Kundell, Jan Ballard, and Katie Bishop—all from the University of Georgia—and Sharon Leathery, of the National Zoological Park's Conservation and Research Center, assisted with accumulation and organization of the bibliographic material and typing of the manuscript. Several people from the University of Georgia Department of Zoology, including Ann Clark, Jeannie Edwards, and Denise Stockton, provided help with various administrative tasks. My supervisors at the University of Georgia (Josh Laerm), Conner Museum at Texas A&I University (Jimmie Picquet), and the Conservation and Research Center (Chris Wemmer) allowed me the flexibility in working hours necessary to complete a task of this type. The following persons provided reviews of various parts of *Nearctic Avian Migrants in the Neotropics* (Rappole et al. 1983), on which a significant portion of the present work is based: Mario Ramos (World Bank); Kevin Winker (Conservation and Research Center); present or former U.S. Fish and Wildlife Service employees Curt Freese, Chandler Robbins, and Mark Shaffer; J. P. Myers (W. Alton Jones Foundation); and Allan Phillips. My coauthors from that work (Gene Morton and Tom Lovejoy of the Smithsonian and Jim Ruos, formerly with the U.S. Fish and Wildlife Service) were also very helpful in this regard. Peter Cannell, science editor for Smithsonian Institution Press, was helpful in keeping the venture on track at a reasonable pace. Rosemary Sheffield, copy editor, helped significantly to improve the style and clarity of the text. Bob Askins, Connecticut College, and Dick Hutto, University of Montana, brought outstanding knowledge of the subject to their thorough and helpful reviews of the manuscript. To all of these people and organizations, I express my heartfelt thanks. Finally, I thank my family, Bonnie, Brigetta, John Jr., and Nathaniel, for their support during the years of work devoted to this enterprise.

1

INTRODUCTION

A significant portion of the avifauna of North America winters in the Neotropics: At least 338 of the 650 or so species that regularly occur north of Mexico migrate south of the tropic of Cancer for the winter (Table 1.1). Additional birds, perhaps as many as 80 to 90 species, breed in the subtropics of southern Texas, Arizona, New Mexico, and northern Mexico and migrate into the Tropics, but their seasonal movements are poorly known (Rappole et al. n.d.). These migrants constitute a highly visible, aesthetically appealing, and biologically significant resource that is economically important for hunting, agriculture, and tourism. This book presents a summary of information and ideas on the biology and conservation of migratory birds, mainly from a Neotropical perspective.

Public and private organizations in the United States and Canada, aware of the need to conserve and enhance avian migrant populations, have funded research, set aside habitat, and attempted to educate the public about the importance of these resources (Bradshaw 1992; Line 1993). However, most Nearctic migrants spend one-half to two-thirds of their life cycle hundreds or even thousands of kilometers away from these good intentions, in Neotropical stopover and wintering areas (Terborgh 1989; Senner 1993).

Knowledge concerning this nonbreeding period of the life cycle is central to an understanding of migrant biology and conservation. There is mounting evidence that the nonbreeding phase is a critical period in terms of migrant survival (Lack 1944, 1968a; Salomonsen 1955; Fretwell 1972; Rappole 1976; Rappole and Warner 1980; Morton and Greenberg 1989; Robbins et al. 1989a; Rappole and McDonald 1994). Yet the abundance, distribution, and ecology of migrants during periods spent away from the breeding area are poorly understood. This problem is especially important in view of the rapid habitat changes currently taking place throughout much of the Neotropics (Myers 1980a; Sader and Joyce 1988; Dirzo and Garcia 1992; Rappole et al. 1994). The synthesis of information that follows represents a step toward a comprehensive understanding of migrants as a valuable, internationally shared natural resource.

Table 1.1

Total number of Nearctic migrant species by family

Family name	Representative members	Number of species
Podicipedidae	Grebes	4
Pelecanidae	Pelicans	2
Phalacrocoracidae	Cormorants	2
Anhingidae	Anhingas	1
Ardeidae	Herons, bitterns	12
Threskiornithidae	Ibises	4
Ciconiidae	Storks	1
Anatidae	Ducks, geese, swans	20
Cathartidae	New World vultures	2
Accipitridae	Hawks, kites, eagles	11
Falconidae	Falcons	4
Rallidae	Rails	7
Gruidae	Cranes	2
Charadriidae	Plovers	8
Haematopodidae	Oystercatchers	1
Recurvirostridae	Avocets	2
Scolopacidae	Sandpipers	30
Laridae	Gulls, terns	21
Columbidae	Pigeons, doves	5
Cuculidae	Cuckoos	3
Strigidae	Owls	3
Caprimulgidae	Goatsuckers	5
Apodidae	Swifts	4
Trochilidae	Hummingbirds	13
Trogonidae	Trogons	1
Alcedinidae	Kingfishers	1
Picidae	Woodpeckers	3
Tyrannidae	Flycatchers	32
Hirundinidae	Swallows	8
Troglodytidae	Wrens	3
Muscicapidae	Thrushes	12
Mimidae	Thrashers	2
Motacillidae	Pipits	2
Bombycillidae	Waxwings	1
Laniidae	Shrikes	1
Vireonidae	Vireos	11
Emberizidae	Warblers, tanagers, orioles, blackbirds	92
Fringillidae	Finches	2
Total		338

In this book the phrase "Nearctic migrant" refers to any nonpelagic Western Hemisphere species, all or part of whose populations breed north of the tropic of Cancer and winter south of that line. There are several Nearctic migrants that do not winter in the Neotropics, but those species are not the focus of this book. Here the principal focus is on that portion of the life cycle spent in the Neotropics, although other significant work relating to the ecology of migrants is included regardless of region of study. The term "Neotropics" refers to that portion of the Western Hemisphere between 23°27' N (tropic of Cancer) and 23°27' S (tropic of Capricorn). Strictly speaking, this definition is not biologically accurate. Nontropical climatic and biogeographic elements extend throughout the Neotropics in mountainous regions, and tropical elements extend as far south as Cape Horn (57° S) (Mayr 1964) and as far north as Bermuda (33° N). Biological definitions of the Tropics vary according to the groups examined, and even within a group there is often little consensus of what constitutes a tropical assemblage. Tramer (1974) cited a shift from tropical to temperate species 350 km south of the tropic of Cancer; Klopfer and MacArthur (1961) located this shift 200 km south of the line; and Gehlbach et al. (1976) concluded that in eastern Mexico the "tropics" begins for birds at the Río Corona, 50 km north of the tropic of Cancer, although they noted that tropical elements extend hundreds of kilometers north into Texas. For the purposes of this book, the exact biogeographic location of the Neotropics is not important. The critical question is, Where do migrants from the north enter communities dominated by species that are resident year-round? For this purpose the tropic of Cancer serves remarkably well (Gehlbach et al. 1976).

Before 1970, interest in the nonbreeding season ecology of migrants was slight among ornithologists. Of course, there was considerable information on distribution and abundance, along with anecdotal bits and pieces buried in general reference works or obscure regional treatments, but with a few notable exceptions (Phillips 1951; Williams 1958; Schwartz 1964; Vogt 1970), little baseline information or analysis relating to nonbreeding ecology of migrants existed.

A Smithsonian Institution symposium published on the topic in 1970 (Buechner and Buechner 1970) shows not only how little work had been done on North American birds in the Tropics but also how little thought had been given to the idea that the nonbreeding portion of the life cycle might be important to migrants. Concern for migratory bird conservation in the face of radical land-use changes in Latin America was a principal focus for William Vogt, who conceived the idea for the conference (Vogt 1970). However, few of the symposium experts addressed the threats to migratory birds, and those who did expressed little apprehension. Roger Tory Peterson, for instance, stated in response to Vogt's presentation: "Our North American migrants, which are marginal when in the tropics, are probably not affected as much as the endemics" (Buechner and Buechner 1970:13).

Since 1970, the amount of data and the number of hypotheses relating to the nonbreeding season ecology of migratory birds have increased exponentially, helped particularly by exciting new insights and information presented in two major symposia: the second Smithsonian Institution symposium on migrants in the Neotropics, held in 1977 at the National Zoological Park's Conservation and Research Center in Front Royal, Virginia (Keast and Morton 1980), and the Manomet Observatory symposium held at Woods Hole, Massachusetts, in 1990 (Hagan and Johnston 1992), whose proceedings were also published by Smithsonian Institution Press. The growing interest in this topic is reflected in the increased number of citations relating to the nonbreeding season biology of migratory birds—from around 3,000 citations in the early 1980s to more than 4,500 at present (Rappole et al. 1983, 1993b). Still, broad gaps in our knowledge remain. Whole groups, such as freshwater wetland and shrub-steppe species, and major topics, such as migrant carrying capacity per habitat, remain virtually uninvestigated.

The main purpose of this work is to synthesize the new information on the nonbreeding season biology of migratory birds in a form that will both encourage interest in the intriguing questions regarding these species and raise public awareness of the potentially damaging effects that current land-use practices in the Neotropics may have on migrants.

I confess that there is an undeniable bias toward passerines in many of the ideas and studies examined here. In part, this bias reflects the primary focus of my own research and that of my students and colleagues. However, it also reflects to a certain extent the amount of available literature. Some regions and groups have been studied much more intensively than others. In any event, the coverage is not intended to suggest that any one group of migrants is more or less important than any other. Excellent treatments of waterfowl migration (Palmer 1975, 1976; Bellrose 1976) and shorebird migration (Morrison 1984; Myers et al. 1984, 1987; Harrington et al. 1986; Morrison and Myers 1989) are available to persons with those particular interests.

The earlier versions of this book (Rappole et al. 1983, 1993b) contain appendixes with common English and Spanish names for Nearctic migrants, patterns of migrant habitat use and food use, maps depicting the Western Hemisphere range for each species, and a summary of the distribution and status of nonbreeding migrants throughout the hemisphere, as well as an annotated bibliography. In this book, these items have been omitted, except for the list of species and some habitat use information.

The book is designed for researchers, conservation biologists, legislators, and anyone with an interest in the biology of migratory birds. The information derives from a selective examination of a vast literature of published and unpublished sources and from the helpful input of more than 200 researchers from various parts of the Western Hemisphere. All facets of migrant non-breeding biology are treated, along with various aspects of conservation. I have excluded discussion of migrant navigation and orientation because they are full topics unto themselves and do not pertain in any special manner to Neotropical migrants.

Chapters are presented on habitat use, resource use, community ecology, migration, evolution, world migration systems, population change, and conservation. These chapters do not serve as complete reviews of these themes, each of which rightly deserves a volume in itself. Rather, I have attempted to address the principal concepts, current thinking, and key unanswered questions in each of these areas, as I see them. My hope is that the information and analysis presented here will stimulate research, conservation, management, and policy-making efforts, and that those efforts will in turn improve understanding and conservation of New World biota in general and Nearctic migrants in particular.

2

HABITAT

IDENTIFICATION OF HABITAT NEEDS

The question of which habitats are used by a species is a seemingly straight-forward one. However, two principal difficulties arise when looking for the answer to that question. First, how are densities of a species accurately assessed across different habitat types, and second, does the number of individuals of a species using a given habitat reflect the species' preference? These difficulties are compounded for migrants, whose preferences may change seasonally and according to their migratory state (Rappole and Warner 1976; Hutto 1981, 1985b).

Survey Methods

Migratory bird habitat use in the Neotropics has been determined in two ways: by capture and by observation. Observational techniques for determining migratory bird habitat use mainly involve audiovisual surveys (Askins et al. 1992; Hutto 1992; Lynch 1992; Powell et al. 1992; Petit et al. 1993). Using these procedures to accurately measure the number of birds in a given habitat is difficult enough on the breeding ground (Conner et al. 1983; Rappole and Waggerman 1986; Rappole et al. 1993a), where researchers have the advantage of males singing to advertise their presence. At stopover areas and on the wintering ground, measuring densities in different habitats is considerably more difficult (Rappole et al. 1992; Rappole, Powell, and Sader, unpub. data). Though some individuals advertise their presence with brief calls, many do not, and vocalization varies widely within species in different habitat types and among species. For instance, Hooded Warblers generally vocalize loudly, regularly, and recognizably throughout the day in tropical habitats, but Wood Thrushes perform 88% of their vocalizations from 0600 to 0900 and from 1600 to 1800 (Rappole and Warner 1980) and are mostly silent during other times of the day. Some birds, like the Yellow-breasted Chat and Yellow-throated Vireo, seldom vocalize at all, at least in the parts of their winter range with which I am familiar, and they are also difficult to detect visually (Rappole, Powell, and Sader, unpub. data). Additionally, birds may vocalize at different rates in different habitat types. Rappole et al. (1994), for example, surveyed for migratory birds in Belize rain forest and second growth using both standardized audiovisual surveys (point counts) and capture surveys, with mist nets placed in a 1-ha grid. In 100 vocalization-dependent point counts, greater numbers of Wood Thrushes were detected in forest than in second growth (13 versus 5 individuals, respectively). These results contrast sharply with standardized netting samples obtained at the same time in the same habitats. In those samples, larger numbers of birds were captured per unit of time in second growth than in forest. In ten 1-ha netting samples, an average of 3.0 birds/250 net-hours were captured in forest versus 4.8 birds/250 net-hours in second growth.

Some studies have used playback of call notes or predator vocalizations to increase detectability of birds on the wintering ground (Lynch 1992). This technique adds another unknown to the equation—namely, variation in responsiveness to playback. Rappole and Warner (1980) found that response to playback varied sharply from species to species and even among individuals within species on wintering sites in rain forest in Veracruz, Mexico. They found that 69% of Hooded Warblers (38 of 55), 46% of Wood Thrushes (6 of 13), and no Wilson's Warblers (0 of 3) or American Redstarts (0 of 2) approached in response to playback of call notes. An additional complication with use of playback is that it is difficult to know what area an individual is being drawn from. Did it come from only a few meters away or from 50 m away? This information is important in estimating species density.

The main method for capturing wintering migrants has been mist netting. This method, too, has its limitations (Karr 1981). Mist nets can be used to sample the avian community from 0 to 2.5 m above the ground, which may be adequate for low habitats or low-foraging species. However, in forested habitats where the canopy can be 35 to 40 m above net height, large portions of the avian community, including many migrants, obviously are missed (Loiselle 1987; Powell et al. 1992; Rappole, Powell, and Sader, unpub. data). Another problem is that members of the same species may forage at different heights in different habitat types, confounding density comparisons among habitats based on net samples.

Despite these difficulties, some researchers have used capture rates in ground-level mist nets to determine "relative abundance" in different habitats for species that are not restricted to the lower strata in habitats with greater canopy heights (Petit et al. 1992; Robbins et al. 1992b). Robbins et al. (1992b) used this procedure to study "relative abundance" of several songbird species in tropical habitats with different structural characteristics and canopy heights, including species known to forage at different heights in different habitats (Magnolia Warbler, American Redstart, Black-and-white Warbler, Northern Parula, Black-throated Blue Warbler, White-eyed Vireo).

Our data comparing foraging heights of Magnolia and Wilson's warblers in different habitats in southern Veracruz and in Belize illustrate the potential biases. These species are common in Veracruz scrub habitats, where they forage from ground level to the top of the canopy, which may be 2 to 3 m high—well within the sampling range of mist nets. Observational data indicate that these species are also common in forests in the same region, though average foraging height is 18.2 m for the Magnolia Warbler and 18.0 m for the Wilson's Warbler—well above net height (Rappole and Warner 1980; Rappole et al. 1992). In a study performed in Belize rain forest (canopy > 10 m) and low second growth (canopy < 3 m), Magnolia Warblers were detected on 28 of 100 point counts in forest and 4 of 100 counts in low second growth. Netting data from ten 1-ha sites sampled in each habitat yielded 0.1 birds/site in forest and 3.6 birds/site in low second growth (Rappole, Powell, and Sader, unpub. data).

Determination of Habitat Preference

In addition to the sampling problems, interpretation of density figures is difficult. Higher density is not necessarily an accurate reflection of a habitat's value from a fitness perspective (Van Horne 1983; Vickery et al. 1992; Winker et al. 1995). One theory predicts that when population size exceeds carrying capacity for optimal habitat, individuals will occupy suboptimal habitats (Brown 1969; Fretwell and Lucas 1970).

Though few tests have been made of habitat use models on the wintering ground, there are some intriguing data on the issue. Rappole and Morton (1985) found higher daily turnover rates in second growth areas for Wilson's and Magnolia warblers than in neighboring rain forest, where "daily turnover rate" is defined as a ratio of the new individuals captured to total number captured per day. Similarly, Rappole et al. (1992) found higher recapture rates per net-hour in second growth than in rain forest for five species of migrants in Veracruz (Least Flycatcher, Gray Catbird, Orange-crowned Warbler, Common Yellowthroat, Yellow-breasted Chat), but eight species showed lower rates in second growth (Yellow-bellied Flycatcher, Wood Thrush, Black-and-white Warbler, Worm-eating Warbler, Ovenbird, Louisiana Waterthrush, Kentucky Warbler, Hooded Warbler). Winker et al. (1995) found that Wood Thrushes, while actually at higher densities in riparian second growth habitat than in adjacent primary rain forest, had higher turnover rates in disturbed riparian habitat than in rain forest. The Wood Thrush populations in rain forest had a higher proportion of individuals holding long-term territories. In a radio-tracking study, Rappole et al. (1989) found that Wood Thrushes originally located and radio-tagged in second growth spent considerable periods during the day investigating Wood Thrush territories in neighboring primary forest.

These findings raise the question of how to distinguish between preferred "optimal" habitat and tolerated "suboptimal" habitat. It may be that in some instances densities can be higher in suboptimal habitats than in optimal habitats because of intraspecific competitive interactions, which force less competitive individuals into marginal environments. This situation may apply to many migratory species under present conditions of habitat alteration in the Neotropics (Robbins et al. 1989a; Rappole and McDonald 1994; Winker et al. 1995).

MIGRANT HABITAT USE

Given the procedural difficulties of measuring and interpreting migrant bird densities accurately, the wide disagreement among biologists about the habitat preferences of migrants at stopover and wintering sites is not surprising. The central question concerning migrant habitat use in the Tropics has focused on migrant ability (or inability) to occupy stable, mature habitats on a long-term basis. Many researchers have reasoned that migrants should be incapable of competing with resident birds for food in these habitats because the migrants are temperate zone species temporarily visiting ecologically packed tropical communities. Therefore migrants should be rare or absent from primary habitats and should concentrate in disturbed areas (MacArthur 1972; Karr 1976; Petit et al. 1993). However, all major New World avian communities include one or more Nearctic migrant species (see regional works in Appendix 6 of Rappole et al. 1983). Though, as discussed above, the habitat preferences of migrants are unclear, every major habitat type that supports avian populations includes migrants. This observation indicates that migrants are not excluded from tropical habitats, primary or otherwise. The habitat distributions for non-breeding Nearctic migrants are given in Rappole et al. (1993b: Appendix 2), based on a thorough review of the basic literature. These data are summarized here in Figure 2.1. Of the 338 species of Nearctic migrants, 106 are found predominantly in aquatic habitats; 120 in desert, grassland, savannah, scrub, or beach habitats; and 112 in forested areas.

Habitat

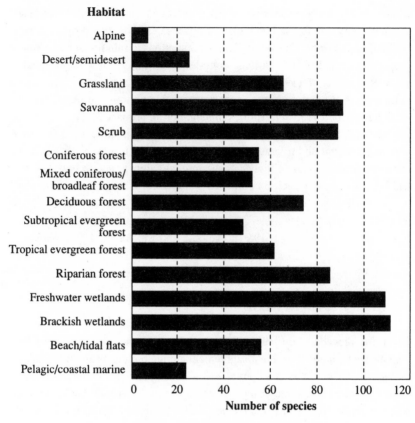

Figure 2.1. Migratory bird habitat use in the Neotropics.

Several recent studies have provided lists of migrant habitat use in the Neotropics. Arendt et al. (1992) gave an exhaustive listing of the habitats in which landbird migrants occur in the Caribbean. Lynch (1989, 1992) provided a similarly thorough listing for the Yucatán Peninsula, as did Blake and Loiselle (1992) and Powell et al. (1992) for northeastern Costa Rica, Hutto (1992) for western Mexico, and Kricher and Davis (1992) and Petit et al. (1992) for Belize.

Forest Habitats

The main debate on habitat use by migrants in the Tropics has focused on forest communities. It is for this habitat that the idea of a hoard of migrants invading a complex, stable, and established resident bird community seems

Table 2.1

Percentage of the avifaunal community composed of Nearctic migrant species for several Neotropical forest types

Forest type	Locality	Percentage	Investigator(s)
Tropical	St. Croix, Lesser Antilles	49	Rappole, unpub. data
Tropical subevergreen	Campeche, Mexico	30–40	Waide 1980
Tropical wet lowland	Veracruz, Mexico	15	Rappole and Warner 1980
	Costa Rica	11–12	Stiles 1980
	Nicaragua	13	Howell 1971
Tropical lower wet montane	Santa Marta, Colombia	4	Johnson 1980
Moist lowland	Canal Zone, Panama	12	Hespenheide 1980
Tropical dry lowland	Costa Rica	22	Stiles 1980
Dry lowland	Canal Zone, Panama	31	Hespenheide 1980
Premontane	Alto Yunda, Colombia	5	Hilty 1980
	Valle del Cauca, Colombia	14	Orejuela et al. 1980
	Costa Rica	8	Stiles 1980
Montane	Costa Rica	6–10	Stiles 1980
Gallery	Colima, Mexico	51	Hutto 1980
Tropical deciduous	Nayarit, Mexico	29	Hutto 1980
Oak woodland	Guadalajara, Mexico	33	Hutto 1980
Pine woodland	Michoacán, Mexico	24	Hutto 1980
Pine-oak woodland	Michoacán, Mexico	29	Hutto 1980
Pine-oak-fir woodland	Michoacán, Mexico	38	Hutto 1980
Amazonia	Peru, Ecuador, Bolivia	6	Pearson 1980

most paradoxical (Karr 1976; Petit et al. 1993), yet nearly a third of all Nearctic migrants accomplish this feat each year. The forests they invade vary from the deciduous forests of the Yucatán Peninsula to the rain forests of the Amazon and the highland pines of the Mexican Cordillera (Table 2.1). In fact, every major forest habitat in the Tropics, including undisturbed wet lowland forests, has Nearctic migrant representatives (Rappole et al. 1983; Loiselle 1987; Powell et al. 1992; Rappole et al. 1992). The number of migrant species in tropical forests varies considerably, depending on forest type and geographic location (Table 2.1), although some tropical wet lowland forests of

South America have the lowest percentage of migrant representation. The implication of these data is clear: Migrants use forest habitats in the Neotropics.

Some investigators have raised the question of whether forest use by migrants is a function of disturbance (Karr 1976; Robinson et al. 1988; Petit et al. 1993, 1995). Fitzpatrick (1980:74), for instance, stated that Eastern Wood-Pewees in Peru were observed "in forest openings along lakeshores and . . . treefall openings." Winker et al. (1990b) found that Wood Thrushes were netted more often in areas within forest that had been subjected to moderate disturbance (tree falls). However, although some migrant species react positively to disturbance, others do not (Lynch 1992; Rappole et al. 1992).

Rappole and Morton (1985) conducted a study comparing abundance and species richness for the migrant and resident community in rain forest before and after severe disturbance. This habitat was undisturbed rain forest with some disturbance around the edge during the initial sampling (1973–1975). When the authors returned in 1980–1981, they found the 6-ha site altered to a mixture of rain forest, edge, pasture, and cornfield (Fig. 2.2). Abundance and species richness had declined for forest-related migrants in this community as well as for residents. Two migrant species disappeared altogether from the later sample, the flock-following Black-and-white and Worm-eating warblers. However, several species of scrub and open country birds were added to the community: Groove-billed Ani (*Crotophaga sulcirostris*), Yellow-rumped Warbler, Melodious Blackbird (*Dives dives*), and Inca Dove (*Columbina inca*).

Second Growth

As noted above, migrants of some species, as well as many resident species, are found in seral stages, or "second growth," of various types. As primary or climax forest disappears in the Neotropics, the use of second growth assumes greater importance (Askins et al. 1992; Robbins et al. 1992b; Staicer 1992). Differences in carrying capacity and long-term suitability of mature versus disturbed habitats have a direct bearing on the survival of many migrants and Neotropical residents. This topic is therefore of major relevance to conservationists and managers concerned with preservation of birds.

Waide (1980) and Lynch (1989, 1992) found that both migrant and resident species used various stages of second growth forest in Yucatán, but species richness in both groups increased with advancing age of the seral stages. The more structurally complex the habitat, the more species of birds that used it, regardless of whether they happened to be endemics or winter residents. Emlen (1977) made a similar discovery in his examination of habitats of varying age and complexity in the Bahamas, as did Rappole and Morton (1985) in southern Veracruz and Askins et al. (1992) in the U.S. Virgin Islands. How-

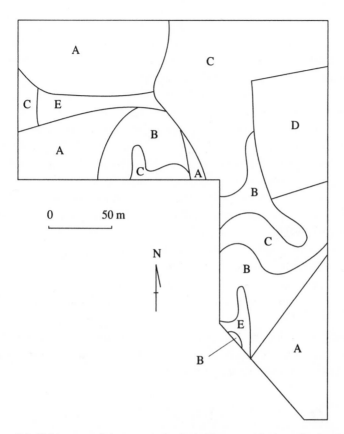

Figure 2.2. Habitats remaining at a study site in Veracruz, Mexico, after modification of the original rain forest. A = open pasture with few large trees; B = forest from which most small to medium-sized trees and undergrowth were cleared; C = rain forest; D = field of mixed cane, corn, bare dirt, scattered brush piles and logs, with a few remaining large trees; E = streamside vegetation, medium and small trees with dense undergrowth. Adapted from Rappole and Morton (1985).

ever, Willis (1980) found a higher number of both migrant and resident species in "young" forest than in "old" forest in Panama.

We propose that the species found in seral stages, whether migratory species or tropical residents, can be divided into four categories: (1) species attracted to the habitat because of its resemblance to mature stages of other habitat types, (2) species attracted because some facets of second growth habitat structure are comparable to those found in mature habitats that are otherwise quite different, (3) species attracted because of temporary resource concentrations, and (4) species adapted to a naturally occurring seral stage, such as riparian flood zone vegetation (Terborgh and Weske 1969). In the first three categories, the word "attracted" may be replaced by the word "forced" in situations where some individuals are prevented from using optimal habitats by conspecifics or by the complete lack of alternatives.

As examples of the first category, Cattle Egrets, anis, Indigo Buntings, Yellow-rumped Warblers, seedeaters, and the like can be found in 1- to 2-year-old second growth rain forest perhaps because at this stage the vegetation resembles a scrub or savannah habitat (Rappole and Warner 1980). Examples of the second category include Lesser Greenlets (*Hylophilus decurtatus*), American Redstarts, Yellow-throated Euphonias (*Euphonia hirundinacea*), and Blue-gray Gnatcatchers. These birds are found in the canopy of mature tropical lowland forest but can also be found in the canopy of early seral stages (Rappole and Warner 1980). Stiles (1980) measured this phenomenon in a variety of forest types in Costa Rica. He found that, for both permanent and winter resident species, "70–95 percent of the birds occurring in the canopy regularly follow the foliage-air interface down along the edges." Examples of the third category, species attracted to seral stages by temporary resource abundance, are discussed in Chapter 3. Examples of the fourth category, species adapted to naturally occurring seral stages, could include species like the Yellow-breasted Chat and Least Flycatcher, though in practice species in categories 1 and 4 might be difficult to distinguish.

Shrub-Steppe, Open, and Aquatic Habitats

As indicated earlier, something less than one-third of all migrants use forests during the nonbreeding seasons, yet most of the debate concerning displacement of migrants by residents has centered on these forest-using species. Little ecological work has been done on aquatic or shrub-steppe species, where "shrub-steppe" is defined to include all habitats with low and/or sparse vegetation—alpine, desert, grassland, savannah, and scrub (see Rappole et al. 1983: Appendixes 2 and 6). In general, the assumption seems to have been that birds in these habitats occupy niches similar to those they occupy in temperate regions.

Actual status (whether migratory or not) of waterbird populations is poorly known in many cases. An examination of the "resident" waterbird faunas of the Panama Canal Zone (Eisenmann and Loftin 1971) revealed that 62% of the species in this assemblage had populations that included Nearctic migrants. Furthermore, band recoveries have shown that permanent and Nearctic winter residents of some of the species coexist in Middle America during the period corresponding to Nearctic winter (Lincoln 1936; Cooke 1938, 1946, 1950; Coffey 1943, 1948). Habitat use comparisons and other relationships between coexisting migrant and resident populations of aquatic species are important research areas that are as yet relatively untouched.

Exclusion of migrants by residents from wetland, scrub, grassland, or desert habitats has not received much interest from New World scientists, although Morel and Bourlière (1962) and Leisler (1990) have made a case for this occurrence in the savannah habitats of Senegal and Kenya, respectively. Presumably, if the hypothesis that "residents" are competitively superior to "migrants" is true, then "residents" should be expected to outcompete "migrants" in any tropical environment where resources are limited, not in just one or a few. If the exclusion phenomenon exists in *some* primary tropical habitats, it seems logical that it should occur in *all* such habitats, perhaps to an even greater or more obvious degree in harsh, open environments because the resident species should be highly adapted to exploiting the meager resources available. Therefore, migrants in naturally occurring shrub-steppe or wetland environments should be subject to the same types of ecological exclusion predicted for lowland rain forest environments, if the phenomenon exists. However, in studies of shorebirds in pampas and beach habitats in Argentina, Myers (1980) found no evidence that migrants were being excluded from the resources of these communities. He observed, "Within both migrants and residents there are spectrums of exploitation systems, each adapted to a critical set of species-specific resources" (Myers 1980:48). He concluded that there is no generalized migrant or resident strategy for resource exploitation. In addition to this evidence, many arid and semiarid zone species of migrants have congeners or conspecifics that are resident members of tropical communities, indicating a long evolutionary association (Rappole et al. 1993b).

SEXUAL DIFFERENCES IN HABITAT USE

Sexual differences in nonbreeding habitat use by birds have been recognized for several decades. Nice (1964) observed that while some adult male Song Sparrows (*Melospiza melodia*) remained on their territories during winter, females, young males, and some adult males departed. Lack (1944) reported a similar situation in the European Robin (*Erithacus rubecula*). Since those early observations, sexual differences in wintering site have been found for a number of short-distance migrants (see reviews by Greenwood 1980, Ketterson and Nolan 1983, and Wunderle 1992). Nisbet and Medway (1972) were the first to report differences in habitat use for long-distance migrants. They found that for Oriental Great Reed Warblers (*Acrocephalus orientalis*) wintering in Malaysia, males outnumbered females in scrub, whereas females were

Table 2.2

Documentation of micro- and macrogeographic differences in wintering site by sex for long-distance Nearctic migrants

Species	Location and investigator(s)
Sayornis phoebe	Texas (Rappole 1988)
Parula americana	Quintana Roo (Lopez Ornat and Greenberg 1990)
Dendroica magnolia	Quintana Roo (Lopez Ornat and Greenberg 1990)
D. caerulescens	Puerto Rico (Wunderle 1992)
D. virens	Quintana Roo (Lopez Ornat and Greenberg 1990)
Mniotilta varia	Veracruz (Rappole 1988)
	Quintana Roo (Lopez Ornat and Greenberg 1990)
Setophaga ruticilla	Quintana Roo (Lopez Ornat and Greenberg 1990)
	Jamaica (Parrish and Sherry 1994)
Oporornis formosus	Veracruz (Rappole 1988)
Wilsonia citrina	Veracruz (Rappole and Warner 1980)
	Yucatán (Lynch et al. 1985; Morton et al. 1987)
	Quintana Roo (Lopez Ornat and Greenberg 1990)
W. pusilla	Veracruz (Rappole 1988)

more numerous in *Phragmites* swamps. Since that report, several authors have documented sexual differences in habitat use for the Hooded Warbler (*Wilsonia citrina*) in Mexico (Rappole and Warner 1980; Lynch et al. 1985; Morton et al. 1987) and for other species elsewhere in the Tropics (Table 2.2).

In a review of differential migration by sex (which we view as a special case of micro- and macrogeographic differences in habitat use by sex), Ketterson and Nolan (1983) proposed five possible explanations for the phenomenon:

1. "Intrasexual competition for breeding resources, which might have caused members of the sex that defends territories to winter nearer the breeding ground" (Fretwell 1972; Myers 1981)
2. "Winter climate, which might have caused members of the smaller-bodied sex to migrate farther toward the south"
3. "Intersexual competition for resources during the non-breeding season, which might have forced members of the subordinate sex to segregate themselves" (Gauthreaux 1978, 1982)
4. "Risk of mortality in transit, which might have varied according to sex and led one sex to abbreviate its migrations"

5. Maximization of "potential reproductive success," which might lead individuals of different sex to migrate to different locations

The last of these explanations, derived from Baker (1978), is not really an explanation. Rather, it is a restatement of the basic premise of the argument—that natural selection governs migration.

Of the remaining explanations, three (1, 2, 4) can be restated as a single hypothesis that could have a much larger number of specific examples:

> *Hypothesis 1:* Males and females have different innate preferences for optimal winter habitat based on differences in morphology, physiology, and/or life history. Micro- and macrogeographic differences in winter distribution result from different choices by members of each sex.

Explanation 3 can be restated as follows:

> *Hypothesis 2:* Males and females have the same innate preferences for optimal winter habitat. Micro- and macrogeographic differences in winter distribution by sex result from male dominance, which forces females into suboptimal environments (Gauthreaux 1978, 1982; Lynch et al. 1985; Parrish and Sherry 1994).

The first hypothesis, that differences in males and females lead to different adaptive responses to the environment is quite attractive in that there appear to be a number of examples in which it is operative. However, some situations remain for which it does not provide an entirely satisfactory explanation (Ketterson and Nolan 1983).

The second hypothesis, that males dominate females, is even less satisfactory as a sole explanation. Certain benefits can be envisioned for females in subordinacy to males during the breeding season, perhaps including protection from forced copulation or deflection of predator focus, particularly if food resources are not limiting. But what would be the fitness benefits during the nonbreeding season, when individuals of the two sexes compete head to head for the same, limiting resources? It is difficult to understand why females would tolerate subordinacy (in an evolutionary sense) unless it entailed some advantage or mitigating factors. If male size or bright coloration provided an advantage in competitive encounters, presumably an intense selection pressure would favor females that demonstrated these traits.

These two hypotheses are not mutually exclusive, and the relative importance of each as an explanation for observed patterns may differ significantly among species. However, tests can evaluate the importance of each hypothesis for a given species.

One such test would be to offer members of each sex of a species a dichotomous choice of habitats and to record their response in a controlled environment. Morton (1990) performed this experiment in the laboratory with Hooded Warblers taken from the nest at 4 to 5 days of age and hand-raised to maturity. Each individual was presented with a choice of two sets of environments in which to forage. The first set was a "tall" (2.29 m) versus a "short" (1.07 m) environment; the second was a "vertical stem" versus an "oblique stem" environment. In each case, the first choice was designed to mimic an aspect of the forest environment in which male Hooded Warblers winter in Yucatán, and the second was designed to mimic the shrubby environment in which females winter. He found that males chose the experimental environments mimicking the forest habitat and females chose those mimicking the shrubby habitat a statistically significant number of times. This result indicates that in nature males and females of this species choose to use the habitats that they occupy in winter, rather than being forced into them. This result was subsequently verified by field tests in Yucatán (Morton et al. 1993).

Several of the species in which sexual differences in winter habitat use have been observed defend solitary feeding territories at a single site for all or part of the winter season. In most instances in which the phenomenon has been described, males are reported in more mature habitats and females are found in the shrubbier habitat (Rappole and Warner 1980; Lynch et al. 1985; Lopez Ornat and Greenberg 1990; Parrish and Sherry 1994). For the species in which this behavior occurs, a second test of the two sexual habitat segregation hypotheses would be to observe whether females move onto territories occupied by males when the male competitors are removed. Morton et al. (1987) performed this experiment with Hooded Warblers in mature forests in Yucatán and found that females on neighboring territories did not take advantage of the absence of male competitors to change habitat types.

A third test of male dominance over females is to present a male competitor to a female on her winter territory and observe her response. This experiment was performed by Rappole and Warner (1980) for Hooded and Magnolia warblers. They found that females of these species defended their territories vigorously against experimentally introduced intruders of either sex. When Stutchbury (1994) performed removal experiments on Hooded Warblers in Yucatán, she found that the sex ratio of replacement birds was similar to that of the original owners, and that females were able to obtain and defend territories in the presence of male floaters (individuals unable to obtain territories).

These findings indicate that, for the Hooded Warbler, sexual differences in habitat selection are a matter of choice (hypothesis 1), at least in part. However, these findings do not rule out an important potential role for malelike plumage. When male Hooded Warblers defend winter territories, they give a series of displays that seem designed to present their black cowl and golden cheek to maximum advantage. Females give the same displays, though fewer than 10% of females have malelike plumage.

The range of variation among species in terms of micro- and macrogeographic sexual differences in use of nonbreeding season habitat is extensive. However, the array of species differences in intersexual variation in morphology and plumage is equally broad. There are species in which both male and female are cryptic but differ somewhat in wing measurements (Eastern Phoebe); species in which males are brightly colored but females are cryptic (Black-throated Blue Warbler); species in which males are brightly colored and females are polymorphic, with some having the appearance of males and others being more cryptic (Northern Oriole) (Rappole 1988); and species in which both sexes are brightly colored (Red-headed Woodpecker, *Melanerpes erythrocephalus*) (Kilham 1978). We propose that the range of variation in plumage and morphology seen between the sexes in different migrant species is related to the range of sexual differences in terms of micro- and macrogeographic habitat use.

Bright (contrasting) plumages and cryptic plumages have fitness costs and benefits that vary according to season and the sex of the individual (Hedrick and Temeles 1989), presenting a fertile field of study for the effects of balancing selection factors (Table 2.3). That such costs exist for bright plumages is reflected by the molt of males of some species (e.g., Scarlet Tanager, Magnolia Warbler, Chestnut-sided Warbler) to more cryptic plumages in fall, forcing an energetically costly prealternate molt in spring to reattain bright breeding plumage, rather than wearing the bright plumage through the nonbreeding period. However, individuals of many species retain bright plumage through the nonbreeding portion of the life cycle despite a possible risk of higher predation rates (but see Baker and Parker 1979 for an alternative explanation). As noted by Butcher and Rohwer (1989:96) in their extensive review of conspicuous coloration in birds, "plumage worn during one season may be primarily an adaptation to another season."

Based on these conflicting factors, the following explanation for the range of intersexual variation in migrant plumages in conjunction with an equally wide range of intersexual variation in habitat use is proposed:

1. We assume that, in the absence of significant sexual selection (Payne 1984), the favored plumage for a bird is cryptic (Rohwer 1975).

Table 2.3

Example costs and benefits imposed by selection forces on individuals wearing bright rather than cryptic plumage

Costs of bright pattern
>   Increased probability of predation for the individual
>   Increased probability of predation for offspring (brightly colored parent bringing attention to nest location or fledgling)
>   Increased aggressive response from territorial males during breeding season
>   Energetic demands of additional molts (for males to molt into a cryptic plumage during nonbreeding season to avoid increased predation due to bright plumage, or for females to molt into a bright plumage to enhance competitive ability)

Benefits of bright pattern
>   Enhanced attractiveness to females during breeding season (males only)
>   Enhanced effectiveness in competition for and defense of breeding territory (males only)
>   Enhanced effectiveness in competition for and defense of nonbreeding season resources (both sexes)

2. In the absence of significant balancing factors, with both sexes competing for the same limiting resources during the nonbreeding season, selection should force them toward equivalent morphology (Rohwer 1975).

3. In many species in which the male and female are identically cryptic, the sexes differ morphologically.

4. Because nonbreeding intersexual competition for resources should force *equivalent* morphology and appearance, the presence of differences could be the result of breeding ground factors (Williamson 1971; Kodric-Brown and Brown 1978; Rappole 1988). One such factor could be a form of "ecological release," in which the male and female members of a mated pair, forced by the nature of their cooperative rearing of young to share a foraging space in which resources are relatively abundant, evolve structural differences to minimize intrapair competition for food and maximize efficiency in feeding of young (Snyder and Wiley 1976; Temeles 1985).

5. The presence of intersexual morphological differences favors the development of macro- or microgeographic differences in foraging habitats (Kodric-Brown and Brown 1978). Such differences could be so great that, under certain circumstances of resource dispersion,

Table 2.4
Outline of a theory of andromimesis to explain malelike plumage in female migratory birds

1. Competition among males for females is intense, favoring evolution of morphological and behavioral characteristics to enhance success in competition for females, resulting in bright plumage patterns in some species (Verner and Willson 1969; Payne 1984).
2. Many of the bright plumage patterns evolved by males are used to enhance agonistic displays (Ficken and Ficken 1962).
3. Several of the same displays used by males in agonistic encounters during the breeding season are used by both sexes in defense of territory during the nonbreeding season (Rappole and Warner 1980).
4. Displays given by females that lack the plumage characters of males are less visually effective. Therefore, selection should favor evolution of malelike characters by females that compete with males for resources during the non breeding season (though Stutchbury [1994] found no indication that malelike coloration in Hooded Warblers enhanced competitive ability).

Source: Based on Rappole (1988).

male-female pairs could coexist on the same nonbreeding site by using different resources or different microhabitats (Zahavi 1971; Leck 1972a; Greenberg and Gradwohl 1980; Wunderle 1992).

6. For species in which the male is brightly patterned, and males compete with females and other males for nonbreeding resources, females should mimic males in nonbreeding appearance (Table 2.4).
7. Costs for malelike plumage (e.g., additional molts) and macro- or microgeographic sexual differences in habitat preferences cause different balances to be struck for different species in terms of morphological resemblance between the sexes.

We propose that the plumages in many species of migrants reflect the balancing effects of opposing selective forces forced by the complexities of fluctuating demands imposed by reproduction, competition, predation, and energetics during different portions of the life cycle. Based on this reasoning, we conclude that there is a relationship between plumage variation and micro- and macrogeographic differences in habitat use between the sexes. We predict that the degree of female mimicry of bright male plumage will vary between species according to the degree of competition for the same optimum non-breeding habitat, as well as a complex suite of additional factors, including intersexual variation in food use, foraging structures, or value in early or late occupation of breeding territory.

AGE DIFFERENCES IN HABITAT USE

Gauthreaux (1982) lumped age and sex together as factors affecting dominance status of an individual with regard to differential migration. However, these two characteristics are quite different in terms of the role of natural selection and the expectations for adaptive response. Natural selection can act to produce different optima for males and females (Slatkin 1984), but it cannot act to change two fundamental competitive liabilities for young birds: (1) the shorter period that selection has acted on them than on adults, so that young birds on average represent less well-adapted individuals than adults, and (2) lack of experience.

Thus, young birds leaving the breeding ground for the first time can be expected to suffer in competition with adults both from lack of experience in terms of locating and exploiting resources and in terms of defending those resources from conspecifics. These differences might lead to a prediction that immature individuals would use foraging sites, strategies, or locations different from those of adults. Immatures might even be predicted to wear plumages different from adult plumages to signal their difference in status and minimize competitive interactions (Rohwer 1975; Morton 1976; Rappole 1983). A few researchers have examined these possibilities in species that form flocks during winter (Fretwell 1969; Rohwer 1975; Ketterson and Nolan 1983). Their work appears to confirm that young birds seek to avoid or mitigate competition with adults in a variety of ways.

Little has been done to examine the competitive relationships between age classes for long-distance migrants that are solitary during the nonbreeding period. Pearson (1980), based on his own observations and those of others, assumed that the different proportions of Summer Tanagers in adult male plumage to those in female or immature plumage in different parts of the species' range indicated that immatures were being forced into less desirable habitat. However, he could not distinguish between adult females and young birds, nor did he provide evidence that members of one plumage type were denying the other preferred resources.

Recent work on the Hooded Warbler by Stutchbury (1994) provides some insight into age class competition for long-distance migrants. In removal experiments on her Mexican study sites, she found that although two-thirds of the original owners of territories were adults, nearly all of the replacements were immatures, raising the possibility that competition for winter territories in this species forces many young birds into floater status or marginal habitats.

Additional work needs to be done to examine the relative habitat needs and competitive relationships between adults and young of the same species for long-distance migrants during the nonbreeding period. A difficulty in performing such studies has been an inability to distinguish young birds from adults accurately once skull pneumatization has begun to reach completion (by December for many passerines). Winker et al. (1994) sought to remedy this problem by developing discriminant functions derived from series of morphological measurements that can be used to determine the sex and age of birds throughout the winter period. These procedures open several exciting possibilities for future investigation.

FRAGMENTATION EFFECTS

The subject of the effects of habitat fragmentation on migratory birds has been thoroughly investigated and reviewed from a breeding ground perspective (Robbins et al. 1989b; Askins et al. 1990; Hagan and Johnston 1992; Rappole and McDonald 1994; DeGraaf and Rappole 1995). Considerably less work has focused on the wintering ground. As reported above, Rappole and Morton (1985) found that the numbers of individuals of forest-related migrants declined when a 6-ha rain forest site in southern Mexico was converted to a mosaic of forest, second growth, crops, and pasture. Two flock-following species of migrants common on the site before fragmentation did not appear in the sample after fragmentation (Black-and-white and Worm-eating warblers). Similarly, Robbins et al. (1992b), in comparing migrant species density and richness in fragmented versus unbroken pieces of similar habitat at several sites in Middle America and the West Indies, found that although some species showed little or no density differences between unbroken and fragmented sites, other species were found in lower numbers or were absent from fragments (e.g., Louisiana Waterthrush and Gray-cheeked Thrush). Askins et al. (1992) examined migrant densities in unbroken tropical forest on the island of St. John in the U.S. Virgin Islands and compared them with densities on small plots of similar habitat on the neighboring island of St. Thomas. They found that, in general, large blocks of habitat on St. John had significantly higher densities of migrants than similar, fragmented habitats on St. Thomas. As in the studies by Rappole and Morton (1985) and Robbins et al. (1992b), they found some species in the fragmented habitats on St. Thomas that were not found on St. John (Black-throated Blue Warbler, Black-throated Green Warbler), and several on the large blocks on St. John that were not found on St. Thomas (Yellow-throated Vireo, Blue-winged Warbler, Magnolia Warbler, Prairie Warbler, Blackpoll Warbler, American Redstart). However, they could not separate matrix (surrounding habitat type) from fragmentation effects, so their results are inconclusive with regard to pinpointing fragmentation effects.

The studies discussed above are among the few that have addressed the potential effects of tropical habitat fragmentation on migrants. These studies indicate that no generalized migrant response exists. Rather, the individuals of each species can be expected to respond according to how their specific needs are affected by the breaking up of habitat. These needs may vary even for the same species in different habitat types or regions. For instance, the Black-and-white Warbler follows individual mixed flocks of birds in Veracruz rain forest, defending the flock itself from intrusion by other Black-and-white Warblers. When this forest type is fragmented into small blocks (<2 ha), the Black-and-white Warbler disappears along with other flock attendants because the flocks have a large home range (>2 ha). In contrast, in Veracruz second growth and in some Caribbean habitats, the Black-and-white Warbler is solitary or associates with small interspecific groups of warblers. Fragmentation of these habitats may have little or no effect on Black-and-white Warbler use of remaining patches.

Species response to fragmentation is likely to vary according to the nature of the surrounding habitats as well. For example, a 1-ha patch of moist forest surrounded by tall second growth may be quite different in terms of its suitability for Wood Thrushes from a 1-ha patch surrounded by heavily grazed pasture.

INTRASEASONAL SHIFTS

Intraseasonal shifts in habitat use have been documented for several species of migrants (Morton 1980; Ramos 1983; Hutto 1985b; Winker et al. 1992d; Parrish and Sherry 1994). However, such shifts are not restricted to Nearctic migrants but occur for many tropical resident species as well. These shifts generally appear to be associated with sharp changes in resource abundance (Ramos 1983). Further review and discussion of the phenomenon are presented in Chapter 3.

USE OF REMOTE SENSING

Remote sensing technology using satellite imagery and aerial photography to determine types, amounts, and rates of change in tropical habitats presents a tool of considerable potential value to biologists interested in the conservation and management of migrants and other species. Green et al. (1987) were the

Table 2.5

Estimated former and current Wood Thrush population sizes on study areas in Costa Rica, Belize, and Mexico

Country and habitat	Estimated former cover (km^2)	Estimated current cover (km^2)	Estimated density	Former population size (thousands)	Current population size (thousands)
Costa Rica[a]					
Forest[b]	1,296.0	310.0	210[c]	272.2	65.1
Second growth	13	578	0	0	0
Open	13	425	0	0	0
Other	—	9	—	—	—
Total	1,322	1,322	—	272.2	65.1
Belize[d]					
Mesic forest[e]	556	416	303	168.5	126.0
Karst forest[f]	238	178	72	17.1	12.8
Second growth	8	79	290	2.3	22.9
Open	8	77	30	0.2	2.3
Other	—	60	—	—	—
Total	810	810	—	188.1	164.0
Mexico[g]					
Forest[h]	538.0	28.0	240	129	6.7
Open	0.4	510.4	0	0	0
Total	538.4	538.4	—	129	6.7

Source: Rappole et al. (1994).

Note: — = status unknown.

a Habitats are based on total area <1,000 m in elevation.
b Forest habitats with canopies >10 m tall located between 1,000 m and 50 m elevation.
c Number of birds/km^2 per 250 net-hours = 3.9.
d Habitats below an elevation of 750 m.
e Seven mesic forest sites were sampled. Number of birds/km^2 per 250 net-hours = 5.6.
f Three karst forest sites were sampled. Number of birds/km^2 per 250 net-hours = 1.3.
g Habitats are based on Landsat habitat data for the San Martín portion of the Tuxtla Mountains, Mexico, from Dirzo and Garcia (1992).
h Defined to include second growth habitats >3 m tall. See text for explanation.

first to use this technology to examine amounts of Hooded Warbler habitat in portions of the Yucatán Peninsula. Since that time, Sader et al. (1991), Powell et al. (1992), and Rappole et al. (1994) have examined migratory bird habitats in wet lowland environments of Costa Rica, Belize, and Mexico.

Obviously, as is the case with any technological tool, some problems and limitations need to be considered before using the tool. The delay can be significant between the time when the satellite records the image data and the time when the scene selection and processing are done to make a coverage usable for field studies. This process can take as long as two to three years, which can make quite a difference in the appearance of early successional or agricultural sites in the Tropics. Habitat identification based on reflectance data can have a significant error rate (30%), especially for younger seral stages. Also, some habitat types that may be significantly different biologically may be difficult to distinguish using remote sensing. Nevertheless, remote sensing combined with aerial photography and accurate ground sampling can provide rapid, areawide assessments of many kinds of available habitats. Most of the problems and difficulties with these studies do not relate to the technology. The chief difficulties, as with other studies of migrant habitat use, lie with development of accurate censusing procedures and determination of preferred habitat. Powell et al. (1992) used a combination of audiovisual censusing and mist netting in which the strengths of each technique helped offset the weaknesses of the other.

When accurate data on species density by habitat are available, remote sensing information can be used to provide regional estimates of population size and rates of change. Green et al. (1987) performed such estimates for the Hooded Warbler, stating that the 2 million ha of forest in the northeastern portion of the Yucatán Peninsula supported roughly 1.2 million Hooded Warblers during the winter period. Similarly, Rappole et al. (1994) used remote sensing along with intensive ground-based studies to obtain estimates of winter populations of Wood Thrushes for southern Veracruz, southern Belize, and northeastern Costa Rica (Table 2.5). They found sharp declines over the past decade in available habitat for Wood Thrushes, particularly in southern Mexico, where forest and second growth below 500 m in elevation has been reduced by 95% to 28 km^2 in the northern portion of the Tuxtla Mountains.

SUMMARY

A concept that continues to persist in analyses of wintering migratory bird habitat use is that migrants possess a "group" habitat use strategy, that is, that migrants have a particular way of relating to their winter habitats that is different from that of nonmigrants using the same habitats. This idea likely will lose its attractiveness as knowledge concerning the extent of subtropical, austral, and intratropical movements among a variety of habitats by many supposedly resident species becomes more widely understood (see Chapter 5). Nevertheless, it is a widely accepted concept at present. For instance, Petit et al. (1995), in their extensive review of habitat use by landbird migrants in the Neotropics, state, "Migratory birds as a group show a stronger affinity toward disturbed habitats than do resident species." In addition, though the authors provide summaries for the habitat preferences of each regional grouping of migrants (Caribbean, northern Central America, and so on), they do not provide a list of the species making up the groups or a list of the specific habitats in which each species has been recorded, further emphasizing a group approach to the question of migrant habitat use.

The chief finding of our review of migratory bird habitat use is that the kinds of habitats used by migrants are as diverse as most other aspects relating to the biology of these species. Migrants are found in all of the major habitat types of the Neotropics. Habitat use is a complex topic, the intricacies of which have only begun to be investigated for migrants during the nonbreeding period. Tantalizing information has been gathered indicating the possibility of sex and age differences in habitat use for some species. These differences, including morphological and behavioral characteristics of males, females, and young individuals, may have profound significance in terms of the evolution of these birds. Migrant habitat use in the Neotropics also has potentially far-reaching conservation implications. New technologies, such as satellite imagery and computer-assisted analysis of topographical databases, make it possible to determine amounts of remaining habitat types and their rates of change. However, the critical question of the relative value (in terms of survival probabilities) of one habitat type versus another remains to be addressed before the importance of the different habitats can be gauged accurately in terms of species conservation.

3

RESOURCE USE

DETERMINATION OF FOOD PREFERENCES

Food use is one of the primary components of an ecological niche and consequently is a vital element in determining the ecological interrelationships of migrant birds. Despite the critical nature of this aspect of avian ecology, food use by migrants is poorly known and understood. The information on food use in this chapter is derived from stomach contents analysis of collected specimens reported in the literature and from observations of feeding birds (Figure 3.1).

There has been a tendency to categorize a given bird species as a consumer of one specific food type—for example, as an insectivore, a frugivore, a carnivore, and so forth. However, as Morton (1971) and others have found, migrants and many other species undergo seasonal changes in their diet. This behavior is common to nearly all mobile species that exist in fluctuating environments. Few species will not take advantage of an easily harvested resource even if they are poorly adapted to compete for it when the resource is scarce.

Appendix 3 in Rappole et al. (1993b) clearly illustrates this point for migrants. Many species, even though apparently specialized morphologically for capturing insects, take a considerable portion of fleshy fruit during their stay in the Tropics. For example, the Eastern Kingbird, with a broad, flat bill adapted for catching flying insects, eats a large amount of fruit during its tropical residence. Furthermore, the schedule of its extensive movements within the Tropics apparently is tied to seasonal patterns of fruit abundance (Morton 1971).

Careful study throughout the course of the nonbreeding period in a variety of habitats and regions will be necessary to develop a complete understanding of the food use patterns of particular migratory species.

SHIFTS IN FOOD PREFERENCES

The resources that a bird is observed to use are not always the same as those that it seems best adapted to use, and frequently an individual may switch

Food type

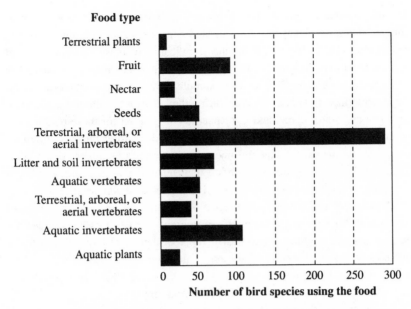

Figure 3.1. Food use by migratory birds in the Neotropics.

from using one category of food to using another. One possible explanation for this behavior is that, in a situation where food is not obviously limiting, individuals may not be forced to focus on harvest of those resources for which they are best adapted. Instead, they may use resources for which they could not compete if the resources were limited, but which they take readily when the resources are abundant. Thus flycatchers may be observed to take fruits when they are plentiful, and seedeaters consume insect larvae under similar circumstances. This type of behavior has been reported for a number of species on the breeding ground (Martin 1987).

Some intraspecific territorial behaviors indicate that food can at times be limiting during migration as well as on the wintering ground. But as is seen in the temperate zone, many parts of the Tropics vary between "boom and bust" cycles. Both migrants and residents may take advantage of the booms, deserting their "normal" niches to skim excesses at reduced energy cost.

Examples of this behavior are found in the literature. Skutch (1980) recorded 95 species of birds that feed on arillate seeds, including many species of normally insectivorous permanent residents and migrants. Similar situations have been reported by other investigators. Morton (1980) reported observing Bay-breasted, Tennessee, and Prothonotary warblers, a Northern Waterthrush, a Royal Flycatcher (*Onychorhynchus coronatus*), a Social Flycatcher (*Myiozetetes similis*), and a Tropical Mockingbird (*Mimus gilvus*) feeding together in sparse, short grass in a cleared pasture where a swarm of midges (Chironomidae) had concentrated near a sewage seep. Other reports by Morton (1972, 1980) support the significance of these observations and may help to explain some of the misconceptions concerning migrant niches and resource use in the Tropics. For example, Bay-breasted Warblers arrive on their wintering grounds in Panama during the latter part of the wet season. Insect resources are scarce, and intraspecific competition for these resources is intense. The birds are often solitary and territorial during this period. However, by late January, fruits and nectar are becoming abundant. Some Bay-breasted Warblers apparently abandon their territories at this time and begin foraging on these abundant resources in loose flocks of conspecifics.

DesGranges and Grant (1980) report a similar situation in a hummingbird community in Colima, Mexico. When migrants arrive, nectar resources are scarce, and the migrants spend nearly 100% of their time on small territories feeding on insects. Later in the season, as flowers become abundant, they and many resident hummingbird species harvest the energetically less costly nectar resources, spending little time in insect foraging.

Calculations of interspecific competition and niche parameters during periods of plenty may have little relevance to factors that actually reflect on the species' place in the community. The ability to capitalize on boom cycles in natural communities is, of course, a part of a species' natural history, and many species of tropical residents as well as migrants capitalize on these local fluctuations (Feinsinger 1980; Myers 1980). Resident species, common in a region of forest during one season, may be completely absent during the next as they track the seasonal variation in resource abundance (Feinsinger 1980; Ramos 1983; Powell and Bjork 1994; Rappole et al. n.d.). How these movements, in either migrants or residents, relate to a species' niche is obscure.

FAT STORAGE

Storage of fat in the form of subcutaneous reserves is an important part of the resource use equation for both migrants and residents. The capability of forming such reserves is critical, not only to the evolution of a migratory habit but also to survival in a competitive environment. In recognition of this importance, fat is often used as a barometer of "body condition" (Bailey 1979; Bain 1980; Johnson et al. 1985; Dufour et al. 1993), implying that fat levels bear a direct relationship to the probability of survival for a given individual. However, the "fat level = body condition" argument ignores the possibility that optimal fat levels for an individual will vary in different environmental situations or for different parts of the annual cycle. Even for transients, this variation is quite clear. What passes for "heavy fat" or "maximum fat load" in one species or population might be considered "moderate" or "low" for another (Nisbet et al. 1963).

Fat storage has been well studied as a specific physiological response to exogenous and endogenous cues prior to migration (Farner 1955; Berthold 1975, 1988; Gwinner 1986). Berthold, Gwinner, and their associates (Gwinner 1990) have established that the physiological fattening responses of migrants are endogenously controlled, genetically programmed to last for the particular period required to complete a migratory journey of a specific distance. However, the capacity to lay down fat reserves does not "turn off" for all individuals of migratory species upon their arrival on the wintering grounds. In fact, there are striking differences between fat levels of sedentary and wandering individuals of the same species wintering in the same region. Our studies of insectivorous migrants wintering in southern Veracruz rain forest have shown that most wintering migrants in apparently good habitat show little subcutaneous fat reserves, but wandering individuals of these same species show a wide range of fat levels, from "none" to "moderate" or even "heavy" fat reserves (Rappole and Warner 1980). Winker (1989) found that average fat levels of wandering Wood Thrushes on wintering grounds in Veracruz were higher than for sedentary conspecifics, even though survival probabilities have been found to be higher among sedentary individuals of this species (Rappole et al. 1989). There are also differences in mean fat levels among species, for both wintering migrants and residents (Table 3.1). Most insectivorous residents, like the Tawny-winged Woodcreeper (*Dendrocincla anabatina*) and Sulphur-rumped Flycatcher (*Myiobius sulphureipygius*), carry little subcutaneous fat. Nevertheless, occasional individuals of these species are captured with moderate or heavy fat reserves (Table 3.1). In contrast, frugivorous species commonly carry moderate fat reserves (Table 3.2).

We have proposed elsewhere that the amount of fat carried by an individual may reflect the availability of resources on which the individual depends (Rappole and Warner 1980). If the individual is using a relatively stable resource pool, such as insects on a fixed territory over a period of months, it can afford to carry relatively low fat reserves. However, if it depends on resources that are scattered and undependable, such as fruits or seeds, the individual should store significant fat reserves to tide it over between locations (in time or space) of one resource pocket and the next (Table 3.2). An alternative explanation, or at least a variation on this theme, is that fat birds are using resources that are more readily converted to fat, such as fruit, whereas lean birds may be using foods that are higher in protein. In any case, the ecology of fat

Table 3.1

Percentage of individuals with moderate or heavy fat for winter and permanent resident species exploiting different food types

Species	N[a]	% individuals with moderate or heavy fat[b]	Food[c]	Residence status[d]
Dendrocincla anabatina	30	6.6	I	P
Xiphorhynchus flavigaster	34	0.0	I	P
Mionectes oleagineus	37	16.2	F,I	P
Myiobius sulphureipygius	36	5.5	I	P
Empidonax flaviventris	22	9.0	I	W
Tityra semifasciata	17	17.6	F,I	P
Pipra mentalis	14	28.5	F,I	P
Henicorhina leucosticta	43	2.3	I	P
Catharus ustulatus	20	10.0	F,I	W
Dumetella carolinensis	49	14.3	F,I	W
Vireo griseus	39	15.3	F,I	W
Seiurus aurocapillus	19	0.0	I	W
Oporornis formosus	31	12.9	I	W
Wilsonia citrina	36	11.1	I	W
W. pusilla	24	0.0	I	W
Basileuterus culicivorus	46	0.0	I	W
Icteria virens	22	13.6	F,I	W
Euphonia hirundinacea	28	10.7	F	P
E. gouldi	25	24.0	F	P
Habia gutturalis	63	7.9	F,I	P
Passerina cyanea	14	21.4	S	W
Sporophila torqueola	27	22.2	S	P

Source: After Rappole and Warner (1980:381).

[a]Number of specimens examined.

[b]Fat classes adapted from Helms and Drury (1960).

[c]Food use classification is based on stomach content analysis of the same birds used in the fat content analysis. F = fruit; F,I = fruit and insects; I = insects; S = seeds. For the F,I category, neither fruits nor insects composed more than 70% of the food in the stomachs examined.

[d]P = permanent resident; W = winter resident.

Table 3.2
Chi-square values for fat class frequencies for avifaunal community members in different food use categories and for migrants versus residents

	Comparison group[a]			
Species food use group	Frugivore/insectivores	Frugivores	Graminivores	Residents[b]
Insectivores	39.24**	16.01**	24.80**	—
Frugivore/insectivores	—	0.70	2.82	—
Frugivores	—	—	0.71	—
Migrants	—	—	—	0.57

Source: Data from Rappole (1976).
[a] — = no comparison made.
[b]Because there are no migrant frugivores, frugivores were omitted from the resident group for this comparison with migrants.
**$p < .01$.

deposition and its relationship to resource availability deserves considerably more attention than it has received to date.

SOCIAL BEHAVIOR AND RESOURCE USE

Territoriality

Several studies have been made of species in which individuals are normally solitary during the nonbreeding period. These studies of the behavior of migrants at stopover and wintering habitats show that numerous species defend transient or winter territories against conspecifics, with territoriality being defined as "any defended area" (Kaufmann 1983). Emlen (1973) suggested that territoriality of wintering birds could be a maladaptive carryover from the breeding ground. However, this observation was made before a thorough examination of the phenomenon had been performed for a number of wintering species. Table 3.3 presents a summary of the studies in which territoriality has been reported for a migrant species during the nonbreeding period. It should be noted that the kinds of evidence on which the reported territoriality is founded are not always strong. Most of the studies base their conclusions on observations of one individual attacking or chasing another, without extensive information on long-term site use by banded individuals. Nevertheless, the data in Table 3.3 illustrate that the commitment to a particular piece of tropical ground by individuals of a number of Nearctic migrant species is a widespread occurrence. It is not restricted to one or a few species in one particular region or to one particular type of bird but is found in many regions, habitat types, and taxonomic groups.

Territoriality is a way of exploiting a certain type of defendable, critical resource. It can be used for short-term defense of an ephemeral resource, such as a nectar source or fruiting tree (Armitage 1955; Kale 1967; Emlen 1973), or for long-term defense of a dispersed, relatively dependable resource, such as bark beetles or soil arthropods. Its long-term, site-specific use as a behavioral tactic by individuals of many species of migrants is important because that use indicates that migrants compete for resources within a wide variety of tropical wetland, forest, scrub, and grassland communities. The energy and time invested in defense of these territories further indicate that the resources defended are critical for the defender's survival (Brown 1964; Kaufmann 1983).

Vocal Displays Used in Territorial Defense

Most of the species that we have investigated include vocalization as part of territorial advertisement. The amount of vocalization varies a great deal among species (Table 3.4) as well as among individuals. Species like the Hooded Warbler (*Wilsonia citrina*) vocalize a great deal, at times approaching 2 minutes per hour, whereas others, like the White-eyed Vireo (*Vireo griseus*), vocalize rarely. The vocalizations used by most species are call notes, that is, short, sharp, species-specific vocalizations. Some species, however, use song occasionally on the wintering ground. For the species that use song, probably both sexes sing. We have confirmed this behavior for a few species by collecting both sexes in the act of singing, including Eastern Phoebe, Least Flycatcher, Yellow-bellied Flycatcher, and White-eyed Vireo (Rappole and Warner 1980; Rappole 1988).

Visual Displays Used in Territorial Defense

Of the species listed in Table 3.3, only a few have been thoroughly investigated in terms of the specifics of nonbreeding season territorial defense. For most, the documentation of territoriality consists of little more than a few observations of one bird chasing or displacing another. For some species, however, considerably more information on the phenomenon has been gathered (Table 3.5). Rappole and Warner (1980) studied a diverse community of wintering migrants in rain forest of southern Veracruz. Most of the individuals of the 21 species of migrants were color-marked for individual recognition. In

Table 3.3
Documentation of intraspecific territoriality in transient (T) and wintering (W) migrants

Species	Locality	Status and investigator(s)
Buteo platypterus	Panama	W (Smith 1980)
Pluvialis squatarola	Argentina	W (Myers et al. 1979)
	California	W (Myers et al. 1979)
P. dominica	Argentina	W (Myers et al. 1979)
Charadrius alexandrinus	California	W (Myers et al. 1979)
C. semipalmatus	California	T (Recher and Recher 1969)
	California	W (Myers et al. 1979)
C. vociferus	California	W (Myers et al. 1979)
Tringa melanoleuca	Argentina	W (Myers et al. 1979)
T. flavipes	Argentina	W (Myers et al. 1979)
T. totanus	Scotland	W (Goss-Custard 1970)
Catoptrophorus	California	W (Myers et al. 1979)
semipalmatus	California	T (Recher and Recher 1969)
Heteroscelus incanus	California	W (Myers et al. 1979)
Actitis macularia	California	W (Myers et al. 1979)
Numenius phaeopus	California	W (Myers et al. 1979)
N. arquata	Scotland	W (Goss-Custard 1970)
Limosa haemastica	Argentina	W (Myers et al. 1979)
Calidris alba	California	T (Recher and Recher 1969)
	California	W (Myers et al. 1979)
	Argentina	W (Myers et al. 1979)
C. pusilla	U.S. Atlantic coast	T (Recher and Recher 1969)
	Manitoba, Canada	T (Hamilton 1959)
C. mauri	California	T (Recher and Recher 1969)
	California	W (Myers et al. 1979)
C. minutilla	California	W (Myers et al. 1979)
C. fuscicollis	Argentina	W (Myers et al. 1979)
C. bairdii	Argentina	W (Myers et al. 1979)
C. melanotos	Argentina	W (Myers et al. 1979)
	Manitoba, Canada	T (Hamilton 1959)
Tryngites subruficollis	Argentina	W (Myers et al. 1979)
Archilochus colubris	Mexico	W (Des Granges and Grant 1980)
A. alexandri	Mexico	W (Des Granges and Grant 1980)
Stellula calliope	Mexico	W (Des Granges and Grant 1980)
Selasphorus rufus	Mexico	W (Des Granges and Grant 1980)
	Arizona	T (Kodric-Brown and Brown 1978)
	California	T (Armitage 1955)

Continued on next page

Table 3.3 (continued)

Species	Locality	Status and investigator(s)
S. sasin	Mexico	W (Des Granges and Grant 1980)
Contopus virens	Peru	W (Fitzpatrick 1980)
Empidonax flaviventris	Mexico	W (Rappole and Warner 1980)
E. virescens	Panama	W (Morton 1980)
E. traillii/alnorum	Panama	W (Gorski 1969)
	Peru	W (Gorski 1971)
E. minimus	Mexico	W (Rappole and Warner 1980)
Sayornis phoebe	Texas	W (Rappole 1988)
Myiarchus crinitus	Panama	W (Morton 1980)
Troglodytes aedon	Texas	W (Rappole, unpub. data)
T. troglodytes	Texas	W (Rappole, unpub. data)
Sylvia melanocephala	Libya	W (Willcox and Willcox 1978)
S. conspicillata	Libya	W (Willcox and Willcox 1978)
Regulus calendula	California	W (Rea 1970)
Ficedula hypoleuca	Portugal	T (Bibby and Green 1980)
Saxicola torquata	Libya	W (Willcox and Willcox 1978)
Myadestes townsendi	California	W (Lederer 1977)
Catharus ustulatus	Colombia	W (Miller 1963)
C. mustelinus	Mexico	W (Rappole and Warner 1980; Rappole et al. 1989; Winker 1989; Winker et al. 1990a)
Dumetella carolinensis	Mexico	W (Rappole and Warner 1980)
Motacilla alba	Israel	W (Zahavi 1971)
Vireo griseus	Mexico	W (Rappole and Warner 1980)
V. flavoviridis	Panama	W (Morton 1980)
Vermivora pinus	Panama	W (Morton 1980)
V. chrysoptera	Panama	W (Morton 1980)
	Costa Rica	W (Powell 1980)
V. celata	Mexico	W (Rappole and Warner 1980)
Parula americana	Puerto Rico	W (Staicer 1992)
Dendroica petechia	Panama	W (Moynihan 1962)
	Mexico	W (Rappole and Warner 1980)
D. pensylvanica	Panama	W (Morton 1980; Greenberg 1984c)
D. magnolia	Mexico	W (Rappole and Warner 1980)
D. tigrina	Puerto Rico	W (Staicer 1992)
D. caerulescens	Puerto Rico	W (Wunderle 1992)
D. virens	Mexico	W (Rappole and Warner 1980)
D. discolor	Jamaica	W (Holmes et al. 1989)

Species	Locality	Status and investigator(s)
D. castanea	Panama	W (Morton 1980; Greenberg 1984c)
Mniotilta varia	Panama	W (Morton 1980)
Setophaga ruticilla	Mexico	W (Rappole and Warner 1980)
	Venezuela	W (Schwartz 1964)
	Jamaica	W (Parrish and Sherry 1994)
Helmitheros vermivorus	Mexico	W (Rappole and Warner 1980)
Seiurus aurocapillus	Mexico	W (Rappole and Warner 1980)
S. noveboracensis	Panama	W (Morton 1980)
	Venezuela	T,W (Schwartz 1963, 1964)
	Texas	T (Rappole and Warner 1976)
S. motacilla	Mexico	W (Rappole and Warner 1980)
	Cuba	W (Eaton 1953)
Oporornis formosus	Mexico	W (Rappole and Warner 1980)
	Panama	W (Morton 1980)
O. philadelphia	Panama	W (Morton 1980)
Geothlypis trichas	Mexico	W (Rappole and Warner 1980)
Wilsonia citrina	Mexico	W (Rappole and Warner 1980)
W. pusilla	Mexico	W (Rappole and Warner 1980)
	Panama	W (Moynihan 1962)
Icteria virens	Mexico	W (Rappole and Warner 1980)
Piranga rubra	Mexico	W (Rappole and Warner 1980)
Icterus spurius	Panama	W (Morton 1980)

addition to recording field observations, the authors attempted to stimulate defensive response by presenting individuals with "intruders"—conspecifics in the form of caged birds, stuffed birds, or mirrors. The level of response to these intruders varied from individual to individual and from species to species. However, individuals of some species responded with stylized visual displays. Reference to the small amount of literature available on breeding ground displays given by the same species indicated that most of the displays were the same as or similar to those used in intense male-male territorial encounters. Dilger (1956b), for instance, reported that the Wood Thrush used "crest raising," "spread," and "wing flick" displays on breeding territory (Fig. 3.2). These same displays were observed on the wintering ground, used by both sexes in territorial encounters (Rappole and Warner 1980). Similarly, the displays used by the American Redstart in defense of territory on the wintering ground conform to those used by male redstarts in male-male encounters on the breeding ground (Ficken and Ficken 1967).

Table 3.4

Mean vocalization time (seconds/hour) for wintering migrants in southern Veracruz rain forest

Species	0600 (4 h)	0700 (6 h)	0800 (3 h)	0900 (4 h)	1000 (4 h)	1100 (4 h)	1200 (2 h)	1300 (5 h)	1400 (6 h)	1500 (4 h)	1600 (8 h)	1700 (5 h)
Empidonax flaviventris	122	23	420	180	195	195	0	0	6	45	25	20
Catharus mustelinus	160	145	80	14	14	50	15	0	5	0	111	330
Vireo griseus	0	3	80	5	5	0	0	0	0	45	10	15
Dendroica magnolia	102	150	100	128	90	110	120	30	40	70	80	30
Oporornis formosus	20	240	60	70	167	133	20	36	27	3	23	104
Wilsonia citrina	42	69	40	66	76	105	25	52	45	13	105	114
W. pusilla	25	215	40	68	25	60	60	0	10	45	108	25
Total	471	845	820	531	572	653	240	118	133	221	462	638

Source: After Rappole and Warner (1980:379).

Note: 0600 = 0600–0659; 0700 = 0700–0759; and so forth. The total number of hours spent listening during the specified period is given in parentheses.

Table 3.5
Evidence for territoriality in wintering Neotropical migrants

Species	Playback response	Periodic calling	Field observation[a]	Intruder experiment[b]	Site tenacity	Homing
Empidonax flaviventris	+	+*#	+	–	+	+
E. minimus	+	+*#	–	+	+	–
Sayornis phoebe	+	+*#	+	–	–	–
Dumetella carolinensis	+	+	+	+	+	–
Catharus mustelinus	+	+	+	+	+	+
Vireo griseus	+	+*#	+	+	+	–
Mniotilta varia	–	+	+	–	+	–
Helmitheros vermivorus	+	+	+	+	+	–

Sources: Sayornis phoebe, Rappole (1988); all others, Rappole and Warner (1980).
*Uses song on occasion.
#Both sexes sing, as confirmed by collection.
[a] Observation of territorial defense involving two free-flying individuals in the field.
[b] "Intruder" experiments involving observation of approach, display, and/or attack by territory owner to a caged conspecific introduced onto its territory. See text.

Figure 3.2. Wood Thrush displays used in defense of winter territory. From Rappole and Warner (1980); drawings by Chris Barkan.

Some differences have been observed between the displays used in male-male encounters on the breeding ground and those used by both sexes in defense of winter territory. For instance, we observed Wood Thrushes using the "upward," "horizontal," and "seesaw" visual displays along with the *zheep* call in territorial encounters on the wintering ground. According to Dilger (1956a,b), these displays, which are used by the *Catharus* thrushes, are not used by the Wood Thrush on the breeding ground. In fact, a major part of Dilger's case for placing the Wood Thrush into a separate genus was based on the absence of these displays in the display repertoire of that species. We see two possible explanations: (1) Dilger missed the displays in his examination of Wood Thrush breeding ground behavior, and (2) some displays that normally are *not* used on the breeding ground *are* used on the wintering ground (Winker and Rappole 1988). The regular use of song in defense of winter territory by females of some species in which the female rarely sings on the breeding ground would seem to support the latter possibility (Winker 1989).

Floaters

Brown (1964) defines territoriality as the defense of a critical resource in short supply. If a resource is in short supply, then its defense must deprive some individuals. Hence, a territorial system, by definition, presupposes the presence of floaters—individuals unable to obtain territories. The presence of floaters has been documented in several populations of wintering migrants in which territoriality is a common behavioral pattern. Schwartz (1964) observed new birds intruding onto the territories of his banded Northern Waterthrushes in December and January in Venezuela. Similarly, de Roo and Deheeger (1969) recorded new, unbanded birds in wintering, banded populations of territorial Great Reed Warblers (*Acrocephalus arundinaceus*) in the Congo, though the authors attributed the presence of these birds to intrusion by previously undiscovered neighbors. Rappole and Warner (1980) documented the presence of floaters on the territories of six species of migrants wintering in southern Veracruz rain forest: Gray Catbird, Magnolia Warbler, Ovenbird, Kentucky Warbler, Wilson's Warbler, and Hooded Warbler. In each case, the floaters were observed from a blind during periods of observation of the territory owners. The usual response to a floater by the territory owner was immediate attack and chase. In two instances, a Hooded Warbler territory owner was removed by shooting. In each instance, the former owner was replaced within a matter of hours by a bird that had been previously observed on the territory as a floater. The change in behavior accompanying the change in status from floater to territory owner was quite marked. As floaters, the birds were quiet, furtive, and immediately submissive to the owner. As territory owners, these former floaters, one male and the other female, vocalized regularly and attacked and chased any intruding conspecifics.

Summary

The foregoing discussion of territoriality as a common behavior associated with migratory bird defense of winter resources should not be read as an assertion that this behavior pattern is the only, or even the most prevalent, resource use strategy among migrants. Of the 338 species of migrants, most have not been examined with regard to their resource use strategies. Also, it should be recognized that territoriality is a behavior that is mediated by specific environmental circumstances, including factors such as density and distribution of prey, and number and status of competitors. So far as we are aware, no species has been shown to be territorial under all resource use situations. As Brown

(1964) has emphasized, territoriality is beneficial only under certain conditions, and having the flexibility to determine the conditions under which territorial behavior is beneficial to the individual is likely to be an important attribute for individuals of most species. Field data indicate that this flexibility is important. Many species reported as territorial in one habitat or situation have been observed in flocks somewhere else. Staicer (1992) reported a range of behavioral responses to conspecifics that varied not only according to species but also according to individuals within a species. She found that Cape May Warblers wintering on her Puerto Rican scrub site were generally territorial; some Northern Parulas were territorial and some were not, even though the latter individuals had distinct home ranges; and Prairie Warblers were secretive and difficult to observe but did not appear to have territories. Yet return rates to winter sites were comparable for all three species (roughly 50%).

Conspecific Flocks

Ketterson and Nolan (1983) summarize information on several studies of short-distance migrants that associate in flocks to use resources scattered in disjunct clumps (seeds and fruits, for instance) during the nonbreeding period in temperate regions. These studies have documented the intricate competitive relationships between different sex and age groups. Unfortunately, no comparable studies of which we are aware have been done on long-distance migrants that normally associate in flocks during the nonbreeding period in the Tropics.

Several species of migrants are seen commonly in flocks of conspecifics during the winter period in the Neotropics. Most of these species have omnivorous, granivorous, or frugivorous food habits—for example, Indigo Bunting, Yellow-rumped Warbler, Northern Oriole, and Eastern Kingbird. They exploit resources that are located in disjunct clumps over space and time, such as fruiting trees and seed-producing weed fields. The competitive dynamics of these flocks have not been explored. Do adult males dominate females and young? Do males, females, and young form different percentages of the flock in different habitats? Are the flocks cohesive? Are members of flocks philopatric to their wintering site? The first study of philopatry of which we are aware was a banding study of Indigo Buntings in Guatemala in which members of this flocking species were recorded returning to the same tropical wintering site from one year to the next (Van Tyne 1932). Little work has been done to date on the ecology and dynamics of conspecific flocks of long-distance migrants on their wintering grounds.

In addition to these species in which conspecific flocks are a regular occurrence, there are reports of individuals forming flocks for species whose members are normally solitary and territorial on the wintering ground. Slud (1964) found that Wood Thrushes, many of whom defend individual feeding territories during the winter (Rappole and Warner 1980; Winker et al. 1990a), occurred on rain forest sites in Costa Rica in small flocks. Rappole and Warner (1980) also observed Wood Thrush flocks of a few birds up to 10 or more, often following prolonged periods of heavy rain.

Some authors have reported loose "collections" of migrants associating with other migrant species (Lack and Lack 1972; Post 1978). Staicer (1992) reported on flocks consisting of Northern Parula and Prairie and Cape May warblers in second growth scrub in Puerto Rico. She found that although many individuals were solitary, others associated in loose flocks with conspecifics as well as with other warbler species. We have observed a similar phenomenon in low (3 m) second growth in Belize, where "associations" including single individuals of Black-and-white Warbler, Wilson's Warbler, and American Redstart were found apparently foraging together through the habitat (Rappole, field notes, 5 Jan. 1992).

Mixed-Species Flocks

The occurrence of species assemblages that forage in fairly cohesive units throughout a remarkably stable home range has been studied in detail by a number of workers (Wagner 1959; Moynihan 1962; Mlecko 1968; Morse 1970; Buskirk 1972; Buskirk et al. 1972; Morton 1973; Willis 1973; Cody 1974; Cruz 1974; Smith 1975; Munn and Terborgh 1979; Gradwohl and Greenberg 1980; Powell 1980, 1985; Munn 1985; Hutto 1988b, 1994). Moynihan (1962), in a classic study, analyzed the structure of interspecific flocks in Panama and found that they generally consisted of species sets that had particular roles within the flock, which he defined as follows:

> *Nuclear species*—a species whose behavior contributes appreciably to stimulate formation and/or maintain cohesion of mixed flocks
> *Passive nuclear species*—a species that is normally joined or followed by individuals of other species more often than its members join or follow individuals of other species
> *Active nuclear species*—a species that normally joins or follows individuals of other species more often than its members are joined or followed by individuals of other species

> *Attendant species*—a species that joins with nuclear species to form flocks but does little to stimulate the formation or maintenance of flock cohesion
> *Regular attendant species*—an attendant species that stays with the flock for long periods of time and is seldom seen away from the flock
> *Occasional attendant species*—an attendant species that does not necessarily remain with the flock for prolonged periods

In these mixed flocks, one or more resident nuclear species, often a family group, is joined by individuals of several other species of both migrants and residents for various periods of time. The resident, nuclear species often defend a territory, and the territory boundary may overlap with that of other nuclear species within the flock and be codefended by them (Munn and Terborgh 1979; Gradwohl and Greenberg 1980). Intraspecific interactions between flock members are rare. Individuals of some migrant and resident species that are normally solitary and territorial will often join a flock while it remains within their territories but will not follow the flock beyond the territory boundaries; these individuals would be categorized as occasional attendants (Moynihan 1962; Gradwohl and Greenberg 1980; Rappole and Warner 1980). Individuals of other species appear to defend the flock itself as a foraging territory, allowing no other members of their species to join and follow the flocks. These individuals would be categorized as regular attendants and include, for example, individuals of Philadelphia Vireo, Yellow-throated Vireo, Black-throated Green Warbler, Black-and-white Warbler, Worm-eating Warbler, Golden-winged Warbler, Blue-winged Warbler, Chestnut-sided Warbler, Summer Tanager, Wilson's Warbler, and Yellow Warbler (Moynihan 1962; Buskirk et al. 1972; Gradwohl and Greenberg 1980; Morton 1980; Rappole and Warner 1980; Greenberg 1984c).

The benefits of this sort of flocking for nuclear species are uncertain. They may be related to kin selection, optimal foraging patterns, reduction in predation, or other undetermined factors (Moynihan 1962; Powell 1985). The benefits to flock joiners and attendants, particularly insectivorous species such as the Black-and-white Warbler, seem largely related to reducing the probability of predation (Buskirk 1972; Willis 1973; Powell 1985). By associating with a group of related birds that have some genetic stake in protecting each other, joiners and attendants receive some of the benefits of flocking at little cost.

A logical question then is, Why are not all migrant insectivores attendants at interspecific flocks instead of only some? The answer may have to do with the amount of area required to support the bird in a particular habitat and the distribution of suitable microhabitat (or resources) within a habitat. If a territory roughly comparable in size to that of a nuclear species' foraging area is required, then the bird should be an attendant at the flock rather than forage alone, because there is little chance that a conspecific could usurp a part of the area and make a living. However, if in this habitat, areas of only 0.2 to 0.3 ha are required to support the bird, or if there are large areas of unsuitable microhabitats between suitable patches, then the disadvantages of wasted foraging time and energy may outweigh the advantages accrued by increased protection from predators.

On the basis of this "resource distribution" explanation for the occurrence of solitary migrants as attendants at foraging flocks, we would predict that, in addition to resource distribution, a bird's tendency to associate with conspecifics or heterospecifics or to forage alone might also depend on the individual's experience and competitive ability (age class, for example) or on habitat type. In some habitats a member of a species might be a continuous flock attendant; in others it might be a joiner, holding a small territory and foraging with the flock only while flock members were on its territory; and in still other habitats the bird might be solitary. As more long-term work on wintering migrant movements and social behavior has been done, evidence of a pattern of changeable behavior in response to different individual and environmental circumstances has begun to accumulate (Table 3.6). Nonbreeding social behavior is seen to be a mutable characteristic, changing according to the needs and abilities of the individual and the dictates of the environment.

ROOST SITES AS A CRITICAL RESOURCE

A resource that often is not considered as critical for avian species is roost sites. We know that individuals will travel long distances between roosting and feeding sites in some species (Ward and Zahavi 1973), indicating the importance of the roost. Possible reasons for the importance of the roost site include (1) reduced vulnerability to predators, (2) centralized locations to obtain information from conspecifics for location of resource concentrations (Ward and Zahavi 1973), and (3) enhanced protection from the weather.

Use of roost sites by normally solitary passerines on their wintering grounds has not been well studied. However, if roost sites are an important limiting resource, evidence of intraspecific competition for roosts among these species might be expected, as well as occasional evidence of individuals

Table 3.6
Reported social behavior of migrants in varying Neotropical habitats and localities

Species	Habitat	Locality	Social system[a]	Investigator(s)
Catharus mustelinus	Rain forest	Mexico	T	Rappole and Warner (1980)
	Rain forest	Mexico	A	Rappole and Warner (1980)
Dendroica petechia	Pasture hedgerow	Mexico	T	Greenberg and Salgado Ortiz (1994)
	Pasture hedgerow	Costa Rica	T	Skutch, in Bent (1953)
	Forest edge	Panama	F	Moynihan (1962)
D. pensylvanica	Rain forest	Panama	F	Greenberg (1984c)
	Rain forest	Costa Rica	F	Powell et al. (1992)
	Pasture hedgerow	Costa Rica	S,T?	Powell et al. (1992)
D. caerulescens	Low scrub	Puerto Rico	A,C,M	Staicer (1992)
	Forest	Jamaica	T	Holmes et al. (1989)
D. virens	Rain forest	Mexico	S,T?	Rappole and Warner (1980)
	Rain forest	El Salvador	C	Dickey and van Rossem (1938)
	Montane forest	Panama	F	Buskirk et al. (1972)
Mniotilta varia	Rain forest	Mexico	F	Rappole and Warner (1980)
	Rain forest	Belize	F	Rappole, pers. obs.
	Pasture hedgerow	Mexico	S	Rappole, pers. obs.
	Low scrub	Puerto Rico	A,C,M	Staicer (1992)
	Second growth	Belize	M	Rappole, pers. obs.
	Montane forest	Panama	J	Buskirk et al. (1972)
	Deciduous forest	Mexico	F	Hutto (1994)
Setophaga ruticilla	Pasture	Panama	A	Morton (1980)
	Low scrub	Puerto Rico	A,C,M	Staicer (1992)
	Forest	Jamaica	T	Holmes et al. (1989)
Wilsonia pusilla	Rain forest	Mexico	T	Rappole and Warner (1980)
	Montane forest	Panama	F	Moynihan (1962)
Piranga rubra	Rain forest edge	Panama	F	Moynihan (1962)
	Rain forest	Mexico	T	Rappole and Warner (1980)
	Deciduous forest	Mexico	F	Hutto (1994)

[a]F = attendant at cohesive interspecific flocks, defended flock against conspecifics; J = joiner of cohesive interspecific flocks, attended for short periods; A = took part in multispecific feeding aggregation at locally abundant food source; C = foraged with flock of conspecifics; M = foraged with flock of individuals of mixed migrant species; T = territorial; S = solitary individual, status unknown.

shifting between a habitat that provides good feeding sites and a different habitat that provides good roosting areas.

Studies of an avian community in Puerto Rico by Staicer (1992) provide some insight into these questions. She describes a situation in which wintering Northern Parula and Prairie and Cape May warblers occupied specific home ranges on a gridded site of low (<10 m) mixed second growth thorn scrub, deciduous forest, and savannah. Each evening during the period of their winter occupancy of the site, however, the birds moved away from this habitat to a roosting habitat characterized by taller, older trees with broader denser canopies than those in the foraging habitat.

Staicer (1992:313) reports:

During the hour before sunset, migrants moved off the grid singly or in small mixed species flocks to roost communally. Movement to and from the roost was qualitatively different from movement within the home range, and was suggestive of migratory flocks, including more rapid movement and use of "tseep" calls among birds moving together. Upon arrival at the roost site, much flying, chasing, and vocalizing ensued as individuals appeared to vie for roost positions. After sunrise migrants moved rapidly back to their home ranges and then engaged in bouts of interspecific "chipping," in which several individuals of migrant and resident warbler species uttered "chips" for periods of ≥1 min.

SUMMARY

Migratory birds use a broad spectrum of foods during migration and on their wintering grounds in the Neotropics. The food types vary not only from species to species but also for individuals within a species, depending on habitat, time of year, and kinds of resources available. Storage of energy in the form of subcutaneous fat reserves is an integral part of the food use equation. The same individual can carry different fat loads, regardless of the amount of food available, depending on the season and on other factors as yet unclear. Social behavior during the nonbreeding season appears to be at least partly a function of the kinds of foods being used. In situations in which food items are relatively evenly dispersed, individuals are often solitary and territorial, defending a space against intrusion by conspecifics. When foods are distributed in disjunct patches, they are often exploited by flocks of conspecifics. The social dynamics of these flocks, which have been thoroughly investigated for temperate, short-distance migrants, have yet to be investigated for long-distance migrants. Under certain circumstances that are not entirely clear, individual migrants join multispecies flocks consisting of resident species and/or other migrants. These flocks can be highly structured in terms of the roles of different individuals within the flock. Migrants often participate as single representatives of their species, defending the entire flock against intrusion by other conspecifics. Roost sites are a potentially critical resource for some migrants but have yet to be studied in any depth.

4

MIGRANTS AS MEMBERS OF TROPICAL COMMUNITIES

The study of tropical avian community ecology has focused primarily on resident species, while the place of the migrant often has been relegated to that of a special form of nonbreeding, fugitive member of these communities and hence has been ignored in community studies. Karr (1971a), for instance, omitted migrants in his comparison of forest and grassland communities in Panama and Illinois, as did MacArthur et al. (1966), Orians (1969), and Howell (1971) in their studies of Neotropical community ecology, even though migrant species composed 10% to 20% of the total number of species for eight months of the year in the communities they studied. In some tropical communities, Nearctic migrants constitute 50% or more of the species and individuals (Hutto 1980; Myers 1980). When the movements of many tropical species that participate in seasonal subtropical and intratropical movements are included within the purview of migration, the phenomenon can be seen as an important, pervasive aspect of many, and perhaps the majority, of tropical avian communities, a phenomenon in which a significant number of species participate (Ramos 1988; Levey and Stiles 1992; Chesser 1994; Powell and Bjork 1994; Rappole et al. n.d.). In this chapter, we address the question, What is the role of migrants in tropical communities, and how does it differ from that of resident species?

THE MIGRANT FORAGING NICHE

The concept of a specific kind of foraging niche for migratory species as opposed to tropical residents has existed for some time (Moreau 1952; Morse 1971). In most instances the descriptions of the migrant niche are presented as an effort to explain what it is about migrants that allows them to coexist with highly specialized tropical residents without actually affecting tropical community function. Moreau (1952) noted that Palearctic migrants in Africa appeared to avoid competition with tropical residents by exploiting superabundant foods. Morel and Bourlière (1962), based on their studies of migrant and resident birds in the savannahs of Senegal, concluded that migrants existed as floating populations *("populations flottantes")* that harvested the excess resources produced in the savannah habitats—resources that could not be tracked by the resident birds, because severe resource shortages occurring at times of the year when migrants are absent limit resident populations.

Willis (1966) expanded this concept to Neotropical communities with his studies of army ant swarms in Panama and elsewhere in Central and South America. He reported on the behavior of 18 species of avian migrants present either as transients or winter residents at the swarms of mainly two ant species: *Eciton burchelli,* which forms large, frequent swarms at regular intervals, and *Labidus praedator,* which forms small, infrequent swarms at irregular intervals. Willis (1966:229) made the following findings:

1. "Migrants are . . . restricted to the periphery of swarms, in poorer areas for foraging, whenever resident species are present."
2. "Resident species definitely restrict the roles of migrant species and exclude them from swarms whenever competition could occur."
3. "Whereas resident birds crowd in to the large and regular swarms of the army ant *Eciton burchelli,* migratory birds are more frequent at the smaller and infrequent swarms of *Labidus praedator.*"

Willis (1966:221) suggested:

Possibly what I have elsewhere called the "irregularity principle" is involved: biological or physical irregularities in the environment create open niches or superabundances of food, because exploitation of a niche always lags behind its appearance. . . . At some times of year or irregularly, low points in availability restrict the populations of any exploiting species, so that at other times there is unused food which a migrant or similarly generalized species can exploit.

Morse (1971) formalized these ideas in his extensive review of the insectivorous bird as an adaptive strategy. He stated, "Migratory species should be most favorably adapted, relatively speaking, to temporary resources and in some cases may never develop adaptations facilitating successful competition for more permanent ones with permanent residents" (Morse 1971:188).

MacArthur (1972) presented a similar view of the role of migratory species in the Tropics, proposing that most migrant species are excluded from residence in tropical communities by one or more "ecological counterparts"—that is, tropical species filling the same or a similar niche. He stated:

Clearly most Texas species can expect to be greeted by a host of ecological counter-parts at the southern end of their [breeding] ranges. Wholly new families of foliage gleaning birds, such as the antbirds (Formicariidae), appear in the tropics. No wonder so few of the American foliage gleaners make it to Panama, and those that do, like the yellow warbler (*Dendroica petechia*), are restricted in Panama by their competitors. The yellow warbler, for example, is restricted [as a breeder] to mangrove swamps. (MacArthur 1972:135)

The implications deriving from this view of tropical community ecology are broad and pervasive, giving rise to a number of conclusions regarding the role of migrants in tropical communities as more or less "fugitive species," that is, species that "are able to survive by their relatively quick and temporary occupancy of suitable new habitats as these first become available" (Mac-Arthur and Wilson 1967:82). Some of these ideas are addressed below.

1. The migrant niches within stable, primary, tropical communities are transient and focus on the exploitation of temporary resource concentrations. This concept is clearly voiced by Brosset (1968:226), based on his studies of tropical forests in Gabon:

The reasons for avoidance of the primary forest by migratory birds must be sought in two directions: 1. Ecologically, the primary forest appears to be a zoologically saturated biotype, where all the ecological niches are occupied by highly adapted species. 2. For psychoethological reasons, it is doubtful whether competition actually occurs. A migratory bird searches for a locality exhibiting a biotope similar to that in which it was born.

The "ecological counterpart" discussed by MacArthur (1972), the concept of migrants as "species that respond positively to disturbance" expressed by Petit et al. (1995), the "floating populations" of Morel and Bourlière (1962), and the "irregularity principle" of Willis (1966) are all important ideas that presume or state this fundamental tenet: Migrants in the Tropics are incapable of occupying stable niches in avian communities because those niches are filled by well-adapted tropical residents; therefore, migrants live on the fringes of communities exploiting temporary resource concentrations. Willis (1966), Leck (1972a–c), Leisler (1990), and others have observed that migrants as a group are subordinate to tropical resident species at resource concentrations. If migrants are dominated in situations where resources are abundant, how could they possibly compete in situations where resources are limited? On this basis, these and other authors have concluded that migrants depend upon such concentrations to survive.

The Wood Thrush was one of the species investigated by Willis (1966) in his influential study of migrant-resident interactions at army ant swarms in

Central America. He found that, as was true for other migrants observed at these swarms, Wood Thrushes were supplanted by resident antbirds. Willis proposed the Black-faced Antthrush (*Formicarius analis*) as a possible resident ecological counterpart to the Wood Thrush.

Subsequent detailed studies of the winter ecology of the Wood Thrush have found that members of this species arrive and establish individual feeding territories roughly 0.5 ha in size in undisturbed, lowland tropical rain forests of Costa Rica, Belize, and Mexico in October (where, incidentally, the Black-faced Antthrush is an uncommon member of the avian community). They remain on these territories until April, when they depart for their temperate breeding areas. In October they return, at rates of 50% to 60%, to the same winter territories they held the previous year. These studies indicate that the Wood Thrush is capable of occupying a feeding niche on a long-term basis in a relatively stable tropical community (Rappole and Warner 1980; Rappole et al. 1989; Winker 1989; Winker et al. 1990a,b; Powell et al. 1992; Rappole et al. 1994).

The Wood Thrush is not the only Nearctic migrant for which such data have been gathered. Similar studies have been performed on several species, including Hooded Warbler (Rappole and Warner 1980; Rappole and Morton 1985), Kentucky Warbler (Karr 1971b; Rappole and Warner 1980; Mabey and Morton 1992), Black-throated Blue Warbler (Holmes and Sherry 1989, 1992; Holmes et al. 1989; Wunderle 1992), American Redstart (Holmes and Sherry 1989, 1992; Holmes et al. 1989; Parrish and Sherry 1994), and Northern Waterthrush (Schwartz 1963, 1964). These studies and others like them have documented the ability of individuals of migrant species to persist on specific sites in complex tropical avian communities throughout the winter period and to return to those sites in subsequent years.

These data indicate that wintering migrants exploit resources in a variety of mature tropical communities, a conclusion reached by a number of researchers who have performed long-term studies of tropical community ecology (Schwartz 1963, 1964, 1980; Lack and Lack 1972; Emlen 1977; Myers 1980; Rappole and Warner 1980; Smith 1980; Stiles 1980).

2. The niches of migrants in the Tropics are preempted by the niches of one or more resident ecological counterparts. The arguments in favor of the existence of ecological counterparts for migrants are principally theoretical. They are necessary from a theoretical perspective to explain the absence of migrants from tropical communities—for example, lowland rain forests. However, as has been made clear above and elsewhere (see Chapter 2), migrants are not absent from these communities.

Research in tropical communities has thus far failed to reveal a migrant-resident ecological counterpart relationship (Chipley 1976; Rappole and Warner 1980). Rather, as evidence has accumulated to indicate that migrants are integral components of most tropical communities, the need (from a theoretical perspective) for an ecological counterpart has disappeared. Along with its disappearance, however, a question has arisen: If migrants have niches in tropical communities, what happens to resources used by migrants during the months when they are absent? Several workers have concluded that the exodus of migrants from tropical communities creates an "ecological vacuum" that no resident species of birds are able to fill (Slud 1960; Chipley 1976; Emlen 1977; Fitzpatrick 1980; Hutto 1980).

In many parts of the Tropics, this ecological vacuum is largely illusory. There is in fact little impetus for residents to try to exploit niches vacated by migrants, because food resources, for many residents, reach their highest levels during the period of migrant absence (Stiles 1980). For these areas, the time of migrant departure coincides with the end of the dry season, when fruits and nectar resources peak, and the beginning of the rainy season, when insect resources peak (Skutch 1950; Stiles 1980; Tramer and Kemp 1980).

Rappole and Warner (1980:386) have described an additional factor acting against the evolution of adaptations by residents to fill migrant niches: "The environment into which a potential competitor would have to evolve is changing radically at roughly six-month intervals. Any permanent resident attempting to capitalize on this situation would have to be adapted to compete in two different environments simultaneously." The theoretical problems associated with such a challenge to adaptation have been explored by Levins (1968). Thus, the concept of an ecological counterpart for migrant species within their winter range seems to have little or no factual or theoretical basis (Stiles 1980). But could counterparts exist at the peripheries of the winter range, preventing further expansion of migrant species into tropical areas? There are no data to answer this question except in cases where range expansion by a migrant species would bring it into competition with a slightly different version of itself (i.e., a closely related congeneric or conspecific population). An examination of the avifauna of the regions bordering migrant ranges argues against the possibility of single-resident counterparts for migrant species. There is no obvious Chestnut-sided Warbler counterpart or Bay-breasted Warbler counterpart, for example, that prevents either of these species from expanding its range southward into South America.

3. Migrants show a high degree of ecological plasticity relative to resident species. If migrants are to fill the role of fugitive species (sensu

MacArthur and Wilson 1967:82), they must possess a series of behavioral adaptations suiting them to that role. Two types of evidence have been put forward to support this idea. First, migrant species on average have been observed to use more winter habitats than resident species (see review in Leisler 1990), and second, migrants have been observed to use a broader range of foraging behaviors (at least in some studies) than residents in the same community (Herrera 1978; Tramer and Kemp 1980).

Two caveats should be noted with regard to the use of number of habitat types occupied by a species as a measure of niche breadth. First, although the use of several habitats could reflect an ability to use a variety of different foraging microhabitats, another explanation could be that this behavior reflects adaptation to a specific microhabitat that is common to a number of different habitat types (e.g., aerial foraging, canopy gleaning, or duff rummaging). Second, overcrowding in the preferred habitat could result in the forced use of marginal habitats by subordinate individuals (Fretwell 1972); in that case, the use of different habitats might reflect the presence of populations exceeding what the preferred habitat can support rather than some innate ability to exploit larger numbers of foraging strata. Making these distinctions can be quite difficult.

A further point regarding the supposed greater niche breadth of migrants versus residents is that migrants as a group are not uniformly generalist in terms of foraging behavior. The few studies that have been done under controlled conditions have shown considerable difference in degree of ecological plasticity for even closely related migrants (Greenberg 1983, 1984a, 1986). Some species, such as Golden-winged and Worm-eating warblers, have a narrow range of foraging behaviors (Greenberg 1986, 1987a,b), whereas others, such as the Yellow-rumped Warbler, appear to utilize a broad range of foraging behaviors (MacArthur 1958; Greenberg 1986).

The work of Powell and Bjork (1994) on the Resplendent Quetzal (*Pharomachrus mocinno*) reveals the shortcomings of broad generalizations about what "migrants" do versus what "residents" do. Powell's group has used radio tracking to demonstrate that the quetzal is a migrant. Although this behavior was long suspected because of the disappearance of this bird from its breeding area for prolonged periods (Wheelwright 1983), Powell's work has confirmed the details of a remarkably precise, seasonal, point-to-point migration by members of this tropical species. Quetzals move 20 to 30 km on a regular, seasonal basis from breeding sites to nonbreeding sites, apparently tracking different cycles of fruit production. Other recent studies indicate that the movements of the quetzal are probably not a unique situation. Many tropical "resident" species may have intratropical migrations (Beebe 1947; McClure

1974; Ramos 1988; Levey and Stiles 1992; Rappole et al. n.d.). If so, then the distinctions drawn between migrants and residents in the past become a matter of degree rather than kind.

4. Migrant morphology is that of a generalist and is much less specialized than the morphology of resident species. Migrants tend to be smaller and have shorter bills, more attenuated wings, and a generally more uniform morphology than residents (Leisler 1990). Herrera (1978:884) has argued that these morphological differences are part of an overall "generalist strategy": "The use of a patchily distributed supply may preclude, or make unnecessary, morphological specializations, so that these species remain closer to the 'morphological standard' (centroid) than resident species."

Not all researchers have accepted this view. For instance, though Cox (1968) found that migrants had decreased degree of culmen length variability and shorter bills on average than residents, he related this finding to interspecific competition and the evolution of migration. For Cox, these differences indicated that some species, when faced with intense interspecific competition, will avoid such competition by migrating rather than by evolution of highly refined foraging structures.

Some procedural questions need to be addressed in each of these studies before we can fully understand their impact. For instance, in a morphological comparison of resident and nonresident birds in Spain by Herrera (1978), the "resident" group was composed of 11 species from six different passerine families. The "migrant" group was composed of 19 species, 13 of which were from one family, Muscicapidae. None of the species compared were congeners. Similarly, Cox (1968:185) used in his comparison "representatives of 20 orders, families, and subfamilies of birds occurring both in North America . . . and in Costa Rica," with no apparent effort to compare closely related species or subspecies. No list of the species used in the analysis was provided.

Both comparisons demonstrate that the assemblage of resident species, as a group, is more diverse than the assemblage of migrants. This fact, however, was never at issue. Clearly, a Central American country such as Panama, with 883 species of birds (Ridgely 1976), will have a greater diversity of morphologies represented than can be expected in the 338 species of Nearctic migrants. If most Nearctic migrants are tropical residents that have evolved the capacity to exploit temporarily abundant resources in the temperate zone, as we and others have proposed (see Chapter 6), then we would predict that the distances to be traversed and the range of temperate habitats available would impose severe filters on the structural and behavioral characteristics of would-be migrants. These filters alone could assure that migrants would likely be more similar to each other than tropical residents unable to exploit distant temperate resources to their reproductive advantage. Short, cambered wings or long, slender bills may be disadvantages for a would-be migrant to the temperate zone (Greenberg 1981a).

The presumption of morphological comparisons of migrant and resident groups is that differences in morphology reflect differences between a migration life history strategy and a resident life history strategy. This presumption is not warranted when a taxonomically disparate group of residents is compared with a group of relatively closely related migrants. In this instance, many factors affect the morphology of the birds in question, besides migration or lack thereof. For instance, when the bill length of an antbird is compared with that of a warbler, seasonal movement strategies are not the only factor being compared. Nor is anything being said with regard to such ecological questions as specialist versus generalist resource use. Rather, different evolutionary histories, foods, competitors, predators, and a myriad of other unknown but potentially important factors are being compared.

It may be that migrants are more generalist than residents or that they do represent the losers in interspecific competitions. However, close relatives should be used to examine such questions. If one or both of these hypotheses are correct, then the expected differences should be visible in comparisons of pairs of Nearctic migrant populations and tropical resident populations of the same species, a large number of which are available for study (see Appendix 1).

5. Migrants are wanderers during their stay in the Tropics because the resources on which they depend are ephemeral. Many studies have documented that migrants are common at ephemeral resource concentrations (see summaries in Karr 1976, Leisler 1990, and Petit et al. 1995). This fact is incontrovertible. What is at question is the meaning of these occurrences.

If migrants were indeed confronted by tropical ecological counterparts in the Tropics, how could they coexist with these species? One mechanism that has been proposed is that migrants could take advantage of easily harvested resources, such as fruit, turning this behavior into a niche of sorts out of necessity (Willis 1966; Morse 1971; Karr 1976; Herrera 1978; Petit et al. 1995). According to this argument, surpluses are continually being produced in different parts of the environment (Feinsinger 1980; Greenberg 1980), and migrants move to track these surpluses.

There is no question that exploitation of ephemeral resources is an important strategy in avian communities. The question is whether or not this is a migrant strategy, that is, a strategy used principally or entirely by most migrants but not by tropical residents. Two lines of evidence indicate that exploitation of ephemeral resources is not a migrant strategy per se. First, migrants are not the only species recorded at temporary resource concentrations. Skutch (1980) recorded 95 species of birds using oil-rich arillate seeds and berries; only 18 species were Nearctic migrants. Leck (1972a) reported 48 species of birds visiting fruiting trees in Panamanian upland and lowland forest and agricultural areas, 18 of which were migrants. McDiarmid et al. (1977) observed 22 species taking fruits of the tropical dry forest tree *Stommadenia donnellsmithii* in Costa Rica, 6 of which were migrants. These and several other studies demonstrate that use of ephemeral resources is a strategy of many species in the Tropics, residents as well as migrants. For most migrants, however, and probably most resident species, this strategy is simply a way to harvest a food source inexpensively. That some resident species are dominant at these resource concentrations may indicate that they are the species that are dependent on the resources and must move continually to track them. As mentioned above, some "resident" species, such as the Resplendent Quetzal, are known to undergo seasonal movements to track fruiting resources (Ramos 1988; Levey and Stiles 1992; Powell and Bjork 1994). Beebe (1947), in the highlands of Venezuela, and McClure (1974), at sites in Malaysia, both captured tropical "resident" species during apparent migratory movements. Beebe (1947) captured lowland resident species at lights at night in a mountain pass at 1,100 m. The birds had heavy fat and empty guts—strong indications of migration.

The second line of evidence indicating that exploitation of ephemeral resources is not necessarily a migrant strategy per se is that nearly all of the migrant species observed at resource concentrations have been found elsewhere to occupy sites in tropical communities on a long-term basis. In fact, the only Nearctic migrant species of which we are aware that seem to use the fugitive strategy almost exclusively and have not been documented at specific tropical sites over extended periods are the Eastern Kingbird and the Cedar Waxwing.

Several banding studies and intensive investigations of the winter ecology of migrants have documented that migrants demonstrate long-term intraseasonal and interseasonal site fidelity in a variety of tropical communities. Nevertheless, there is no question that migrants, like other species, exploit resource concentrations. Many authors have described multispecies congregations at resources as diverse as midge swarms (Morton 1980), ant raids (Willis 1966), monarch butterfly roosts (Calvert et al. 1979), fruiting trees (Leck 1972a–c; Hilty 1980), flowering plants (DesGranges and Grant 1980; Feinsinger 1980), burning fields (Chapin 1932; Leisler 1990), garbage dumps (Burger et al. 1980), fish kills, bird feeders, and so forth. Some workers have even described in detail the levels of intraspecific and interspecific competitive interaction that occur at these concentrations, apparently assuming that the squabbling at these normally ephemeral or artificial resources was the most important aspect of the niche for the species involved. For most migrants, this assumption is not true. The migrant catbirds, thrushes, vireos, flycatchers, and warblers and even many of the resident species (Howe and De Steven 1979; Skutch 1980) that use fruits as a resource are not all obligate frugivores, nor are they all wanderers.

It has been assumed that when the migrant left the fruiting tree, ant swarm, or flower patch, it left for some other ephemeral resource (Morse 1971). This assumption has been found to be correct for transients passing through, or floaters—that is, individuals unable to compete with conspecifics for stable resources in the community (Rappole and Warner 1980; Rappole et al. 1989). However, long-term banding studies have shown that members of many migrant and resident species, including many of those attending resource concentrations, have distinct home ranges and show site fidelity throughout a winter season (Rappole and Warner 1980; Rappole et al. 1992; Staicer 1992) and from one season to the next (Table 4.1).

In some parts of the Tropics, migrants arrive when insect resources appear to be at their peak (Hutto 1980; Pearson 1980). In parts of the Middle and South American Tropics, however, migrants arrive at the end of the rainy season when insect resources are steadily declining toward lows that are reached and maintained during most of the wintering period (Hespenheide 1975; Emlen 1977; Buskirk 1980; Janzen 1980; Morton 1980; Orejuela et al. 1980; Stiles 1980; Waide 1980; Hutto 1986). Thus, many migrants join tropical communities when competition for resources is most intense. Young produced by the endemics and swarms of incoming adult and young migrants are competing for dwindling or, at best, stable resources. If migrants were not adapted to compete using stable niches in tropical communities, they could not survive. Even the temporary resource flushes would be exhausted by wandering young of resident species forced out of their normal habitats and niches by population pressure at this critical time of the year. That such wanderings of even the most sedentary species of tropical residents occur is well documented (Fogden 1972; Willis 1974).

Some of the findings of long-term banding and radio-tracking studies are relevant to the question of whether migrants, as a group, are wanderers. From 1982 to 1988, we banded 335 Wood Thrushes in lowland rain forest and neighboring 1- to 5-year-old second growth in southern Veracruz Mexico (Rappole et al. 1989). Sixty-one birds were fitted with radio transmitters. Of those 61, we found that 34 birds were sedentary throughout the life of the transmitter (average = 13.2 days); that is, they moved less than 150 m from point of capture. All of the sedentary birds were located on territories in primary forest. Twenty-seven birds were wanderers; that is, they moved more

Table 4.1

Nearctic migrants showing site fidelity to winter quarters

Species	Locality and investigator(s)
Anas discors	Panama (Briggs 1977)
Charadrius semipalmatus	Costa Rica (Stiles and Smith 1977)
Actitis macularia	Costa Rica (Stiles and Smith 1977)
Calidris pusilla	Costa Rica (Stiles and Smith 1977)
C. mauri	Costa Rica (Stiles and Smith 1977)
Limnodromus griseus	Costa Rica (Stiles and Smith 1977)
Gallinago gallinago	Texas (Arnold 1981)
Sphyrapicus varius	Belize (Nickell 1968)
Contopus borealis	Belize (Nickell 1968)
Empidonax flaviventris	Mexico (Ely et al. 1977)
E. minimus	Mexico (Ely et al. 1977; Rappole and Warner 1980)
Myiarchus crinitus	Belize (Nickell 1968)
	Mexico (Ely 1973)
Stelgidopteryx serripennis	Belize (Nickell 1968)
Regulus calendula	California (Rea 1970)
Catharus mustelinus	Mexico (Rappole and Warner 1980; Winker 1989; Winker et al. 1990a)
	Costa Rica (Blake and Loiselle 1992)
Dumetella carolinensis	Belize (Nickell 1968)
	Mexico (Ely et al. 1977)
Vireo griseus	Mexico (Ely et al. 1977)
V. solitarius	Mexico (Ely et al. 1977)
Vermivora peregrina	Panama (Loftin et al. 1967)
	El Salvador (Thurber and Villeda C. 1976)
Parula americana	Jamaica (Diamond and Smith 1973)
	Hispaniola (Woods 1975)
	Puerto Rico (Staicer 1992)
Dendroica petechia	Belize (Nickell 1968)
	El Salvador (Thurber and Villeda C. 1976)
D. pensylvanica	Panama (Loftin et al. 1967)
D. magnolia	Belize (Nickell 1968)
	El Salvador (Thurber and Villeda C. 1976)
	Mexico (Ely et al. 1977; Rappole and Warner 1980)
D. tigrina	Hispaniola (Woods 1975)
	Puerto Rico (Staicer 1992)
D. caerulescens	Jamaica (Diamond and Smith 1973; Holmes and Sherry 1992)
	Hispaniola (Woods 1975)

Continued on next page

Table 4.1 **(continued)**

Species	Locality and investigator(s)
D. townsendi	El Salvador (Thurber and Villeda C. 1972)
D. virens	Mexico (Ely et al. 1977)
D. discolor	Jamaica (Diamond and Smith 1973)
	Hispaniola (Woods 1975)
	Puerto Rico (Faaborg and Winters 1979; Staicer 1992)
D. palmarum	Hispaniola (Woods 1975)
Mniotilta varia	Panama (Loftin et al. 1967)
	El Salvador (Thurber and Villeda C. 1972)
	Jamaica (Diamond and Smith 1973)
	Hispaniola (Woods 1975)
	Mexico (Ely et al. 1977)
	Puerto Rico (Faaborg and Winters 1979)
Setophaga ruticilla	Belize (Nickell 1968)
	Hispaniola (Woods 1975)
	Puerto Rico (Faaborg and Winters 1979)
	Jamaica (Holmes and Sherry 1992)
Helmitheros vermivorus	Panama (Loftin et al. 1967)
	Jamaica (Diamond and Smith 1973)
	Hispaniola (Woods 1975)
	Mexico (Ely et al. 1977)
Limnothlypis swainsonii	Jamaica (Diamond and Smith 1973)
Seiurus aurocapillus	Panama (Loftin et al. 1967)
	El Salvador (Thurber and Villeda C. 1972)
	Jamaica (Diamond and Smith 1973)
	Hispaniola (Woods 1975)
	Mexico (Ely et al. 1977; Rappole and Warner 1980)
	Puerto Rico (Faaborg and Winters 1979)
	Costa Rica (Blake and Loiselle 1992)
S. noveboracensis	Trinidad (Snow and Snow 1960)
	Venezuela (Schwartz 1964)
	Panama (Loftin et al. 1967)
	Belize (Nickell 1968)
	Jamaica (Diamond and Smith 1973)
S. motacilla	Mexico (Rappole and Warner 1980)
Oporornis formosus	Panama (Loftin et al. 1967; Mabey and Morton 1992)
	Mexico (Ely et al. 1977; Rappole and Warner 1980)
	Costa Rica (Blake and Loiselle 1992)

Species	Locality and investigator(s)
O. philadelphia	Panama (Loftin et al. 1967)
O. tolmiei	El Salvador (Thurber and Villeda C. 1976)
	Mexico (Ely et al. 1977)
Geothlypis trichas	Belize (Nickell 1968)
	Hispaniola (Woods 1975)
	El Salvador (Thurber and Villeda C. 1976)
Wilsonia citrina	Belize (Nickell 1968)
	Mexico (Ely et al. 1977; Rappole and Warner 1980)
W. pusilla	Belize (Nickell 1968)
	El Salvador (Thurber and Villeda C. 1976)
	Mexico (Ely et al. 1977; Rappole and Warner 1980)
W. canadensis	Panama (Loftin 1977)
Icteria virens	Belize (Nickell 1968)
	El Salvador (Thurber and Villeda C. 1976)
	Mexico (Ely et al. 1977)
	Panama (Loftin 1977)
Piranga rubra	Belize (Nickell 1968)
	Panama (Loftin 1977)
Pheucticus ludovicianus	Belize (Nickell 1968)
Guiraca caerulea	Belize (Nickell 1968)
Passerina cyanea	Guatemala (Van Tyne 1932)
	Belize (Nickell 1968)
P. ciris	El Salvador (Thurber and Villeda C. 1976)
	Mexico (Ely et al. 1977)
Icterus spurius	Belize (Nickell 1968)
I. galbula	Panama (Loftin 1977)

than 150 m from point of capture (average transmitter life = 9.9 days). The wanderers demonstrated a number of remarkable behaviors. Some remained relatively sedentary in patches of second growth for several days, then moved to another patch. Others joined a flock of birds at an army ant swarm in primary forest for several days before moving elsewhere. Two birds roosted and spent the early mornings in a particular patch of second growth but spent most of the day moving considerable distances (several hundred meters) through neighboring primary forest. The movements of Wood Thrush BS 381 over a 7-day period in January 1988 are shown in Figure 4.1. Four birds that began the study as wanderers in second growth eventually settled on territories in primary forest and became sedentary. No radio-tracked sedentary birds became wanderers. During the course of the study, 2 sedentary birds and 7 wan-

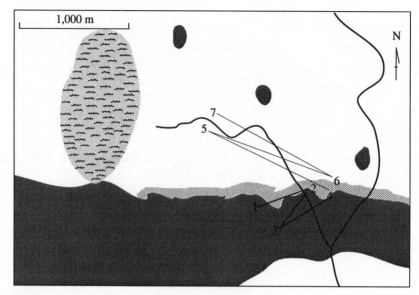

Figure 4.1. Movements of Wood Thrush BS 381. Detection points are numbered consecutively (1 to 7) from capture and release to last detection. The oval area at left is a lake. Areas shaded in dark gray represent rain forest. Areas shaded in light gray represent second growth (3–10 m in canopy height). Blank areas are pasture or low second growth (<1 m). The two heavy lines forming a Y on the right represent roads through the study area. Adapted from Rappole et al. (1989).

derers died, demonstrating a significant difference in daily survival rate between the two groups (sedentary = 0.996, wanderer = 0.974). All deaths were the result of mammalian or avian predation.

These findings indicate that membership in the wanderer class or the sedentary class for Wood Thrushes was probably dependent on amount of available preferred winter habitat. Those birds that were able to locate and defend a territory in primary habitat settled, and those that were not able to do so continued to move, sometimes over distances of several kilometers (Rappole et al. 1989; Winker 1989; Winker et al. 1990a,b). Currently, little of this habitat remains in the region; less than 5% of areas that were covered with lowland rain forest (<500 m) at the time of European settlement still has remnants of this habitat (Dirzo and Garcia 1992; Rappole et al. 1994).

6. Migrants are restricted to communities in marginal habitats or marginal parts of habitats in the Tropics. Many studies have documented that migrants occur at higher densities in second growth and other disturbed areas than in tropical primary habitats (Petit et al. 1995). However, a number of other studies have shown that migrants occur in all major primary habitats of the Tropics as well (Chapter 2; Rappole et al. 1993b: Appendix 2). Based on numerous banding studies and surveys, both extensive and intensive in scope, it is clear that migrants are not restricted to marginal or secondary habitats. Nevertheless, they have been recorded at high densities on a number of disturbed sites (Karr 1976; Hutto 1980, 1989, 1992; Petit et al. 1995).

One possible explanation for this occurrence is that, for some species—for example, Least Flycatcher and Yellow-breasted Chat—a scrub environment of low, dense thickets such as many types of second growth is actually a preferred habitat type. The work of Terborgh and Weske (1969) in Peru has shown that some forms of second growth provide a number of microhabitats similar to those found in other habitat types. Thus, some lowland forest canopy species occur in second growth canopy (Stiles 1980), and species found in thicket or thorn forest habitats can be found in dense second growth as well.

A second explanation for the occurrence of migrants in second growth in greater numbers than residents derives from the theoretical predictions of Fretwell (1972) regarding the effects of overcrowding on habitat suitability. Fretwell proposed that as a habitat fills with individuals, the basic suitability of that habitat declines until it equals that of marginal habitats. When this occurs, the marginal habitats will begin to fill with birds as well. If migrant breeding habitat allows more individuals to be produced than can be supported in the available winter habitat with highest basic suitability, then the excess individuals should spill over into marginal habitats.

SEASONAL CHANGE IN MIGRANT FORAGING NICHE BREADTH

Only within the last two decades or so have studies been done to determine the characteristics of the niches of individual migrant species in the Neotropics (Baker and Baker 1973; Wiedenfeld 1992). Pioneering studies by Chipley (1976), Rabenold (1980), and Bennett (1980) signaled an awakening interest in the ecological complexities of species that experience such dramatic shifts in habitat use during the course of an annual cycle. Each of these workers studied members of the wood warbler subfamily, Parulinae. They compared niche breadth on the wintering ground with that on the breeding ground and then formulated theories of migrant ecology to explain their separate findings. The conflicting results and conclusions from the three studies are summarized in Table 4.2.

Table 4.2
Results of three studies on migrant wood warblers comparing breadths of wintering and breeding ground niches

Species	Investigator	Breadth in winter	Breadth in summer	Explanation
Dendroica fusca	Chipley (1976)	Broad	Narrow	The presence of *fewer* competitors on the wintering ground allows members of this species to use *more* foraging strategies.
D. virens	Rabenold (1980)	Broad	Narrow	The presence of *more* competitors on the wintering ground forces members of this species to use *more* foraging strategies.
Setophaga ruticilla	Bennett (1980)	Narrow	Broad	The presence of *more* competitors on the wintering ground forces members of this species to use *fewer* foraging strategies.

These three niche studies illustrate the kinds of problems that plague many studies of this type, particularly those involving migrants—namely, how is the niche to be measured, and when the chosen parameters have been measured, what do the numbers say about the way the animal lives? If the animal is not in a situation where resources are limiting, then the observed foraging behaviors may have little to do with niche structure; or the parameters chosen for measurement may not accurately reflect the important variables in the species' foraging niche. Two scenarios illustrating these problems as they pertain to migrants are given below.

Scenario 1. Assume that species A is adapted to glean green insects from leaves. It has inherited several behaviors evolved over generations of natural selection that enable it to be very good at finding and capturing prey. These behaviors include stealthy approach; stretching, peering, and careful examination of leaf surfaces; quick gleaning; and a swift flycatching movement if the insect flies during the bird's site examination. If in habitat X, mayflies (Ephemeroptera) are abundant, the bird casts aside its entire range of foraging behaviors except that which is appropriate for obtaining the abundant prey—in this case, flycatching. Or perhaps inchworms (Lepidoptera: Geometridae) are abundant, so the bird uses only the gleaning movement and forgoes its usual stalking, peering, and flycatching behaviors.

In contrast to habitat X, where food is abundant, food is scarce in habitat Y. Intraspecific and interspecific competition for these resources is intense. Individuals of species A use the entire range of foraging behaviors and techniques in this situation. They must do so, because these are the special methods that they can use better than any other species.

An ecologist observing the bird in habitat X might conclude that it had a narrow range of foraging behaviors. The same ecologist observing a member of the same species in habitat Y might conclude that the species had a wide range of foraging behaviors. These data might then lead the ecologist to conclude that the bird had a broad niche in habitat Y and a narrow niche in habitat X, even though the comparison has nothing to do with niche breadth but merely reflects the bird's reaction to different levels of resource availability.

Scenario 2. Suppose that the foraging niche of species B involves exploring clumps of dead leaves for arthropod eggs and larvae. In habitat X, 90% of these leaf clumps occur on the forest floor, and the bird accordingly spends 90% of its foraging time on the forest floor. In habitat Y, clumps of dead leaves are scattered through several levels of the forest because the forest is evergreen, the leaves are larger, and there are more intermediate-sized trees. Members of species B spend equal amounts of time at a number of different levels in the forest.

An ecologist observing species B in both habitats, and using foraging height as an important parameter of niche breadth, might conclude that species B had a broader niche in habitat Y than in habitat X, although in this case comparative foraging height is relatively unimportant to the bird's niche breadth.

The purpose of these examples is to illustrate some of the formidable difficulties that can be expected in comparing niche breadth for a species during different stages of its annual cycle. Marked variations occur from one seasonally occupied site to the next, not only in the number of interspecific competitors but also in the prey base, the structure of the habitat, the potential predators, and a variety of other factors difficult to measure or control. Problems of this type may be responsible for the discrepancies in findings and interpretations in the investigations summarized in Table 4.2 and other studies of comparative niche breadth that lack firm knowledge of whether resources are limiting and how resources might be distributed in the widely divergent habitats compared. If resources are not limiting or relative distribution of resources is not understood, then the niche parameters measured, and the comparisons and resultant speculations regarding ecological release and other reflections on levels of competition in the different habitats, are of little value (Wiens 1977).

NEOPHOBIA

Neophobia is the fear of feeding on new foods or approaching new situations (Barnett 1958; Greenberg 1990). A clever set of laboratory experiments with several Nearctic migrant wood warblers has shown apparent differences in amounts of neophobia among species (Greenberg 1983, 1984a,b, 1990), which in turn may affect the relative amounts of ecological plasticity these species demonstrate when confronted with variable environments under natural conditions. The amount of neophobia an individual shows has both innate and learned components; older individuals may show less neophobia than young birds, depending on their prior experience. Greenberg (1983) hypothesized that fundamental differences among major groups of species may derive from subtle variations in amounts of innate neophobia expressed by members of each group. In particular, the wood warbler (Parulinae) genus *Dendroica* has 21 species that breed in the temperate zone, 7 of which breed primarily in deciduous forest or second growth, and 14 primarily in coniferous forest. Morphologically, the species are quite similar. Based on the results of his experiments and his observations of the birds in the field, Greenberg (1983:450) hypothesized as to why coniferous-breeding *Dendroica* (e.g., Bay-breasted, Magnolia, and Yellow-rumped warblers) have less neophobia than deciduous-breeding members of the genus (e.g., Chestnut-sided and Black-throated Blue warblers): "As coniferous forest breeding birds, they are less well adapted for broad-leaved insectivory during the winter. Their increased plasticity allows them to exploit a wider range of possible resources and capitalize on shifts in insect abundance."

A difficulty with the hypothesis is that ecological plasticity potentially has innate foraging behavior and morphological components in addition to those attributable to neophobia, and designing experiments that eliminate these other factors or hold them constant is problematic. Greenberg's experiments have demonstrated that members of closely related species exhibit marked differences in their reaction to food items presented in novel foraging sites. However, it is difficult to rule out the possible effects of interspecific differences in morphology or foraging behavior.

A second question raised by this hypothesis concerns the relative role of selection occurring at different times of the annual cycle on migrant neophobia, foraging behavior, and morphology. Is breeding season habitat, nonbreeding habitat, or some combination of habitats most important in shaping these adaptations?

An obvious extension of Greenberg's hypothesis would be to predict that migrants as a group have less neophobia and greater ecological plasticity than residents (Leisler 1990)—a prediction that is very similar to the ideas presented by Willis (1966) and Morse (1971), with the added attraction of providing an apparently labile psychological component on which selection could act in a relatively short period. A possible test of this prediction would be to compare the variation in ecological plasticity for populations of the same species that (1) breed in temperate coniferous forest and winter in broadleaf evergreen forest, (2) breed in temperate deciduous forest and winter in broadleaf evergreen forest, and (3) are resident year-round in broadleaf evergreen forest in the Tropics. The Red-eyed Vireo is such a species. It has breeding populations that nest in temperate deciduous forest in the eastern United States and winter in the Amazon basin, other breeding populations that nest in the coniferous forests of northeastern United States and Canada and winter in the Amazon, and a close relative (conspecific?) that is resident in the Amazon, *Vireo (olivaceus) chivi*. If wintering ground habitats dictate degree of ecological plasticity, then the foraging behaviors of members of these three populations should be little or no different. However, if breeding ground habitats are critical, then the behaviors should be markedly different, with the coniferous forest–breeding populations showing the greatest plasticity and the tropical resident populations showing the least. If differences are found, neophobia would be a likely source because, as Greenberg (1990) has pointed out, changes in relative amounts of neophobia are potentially quite rapid in an evolutionary sense. Additionally, they can also be affected by learning.

MIGRANTS VERSUS RESIDENTS IN TROPICAL COMMUNITIES

Direct Migrant-Resident Interaction

A number of studies have documented direct migrant-resident interactions in which a resident species actually attacked a migrant (Willis 1966; DesGranges and Grant 1980; Hilty 1980; Johnson 1980). Most of these reports come from observations at resource clumps (such as ant swarms, fruiting trees, or flower patches), situations in which the competition is related to the niches of the participants in complex and poorly understood ways.

Studies that concentrate on observing the day-to-day activities of migrant individuals in primary habitat types away from resource concentrations report few interspecific agonistic interactions of any sort, though intraspecific agonistic interactions are often common, particularly just after arrival on wintering sites (Schwartz 1964). Fewer than 10 interspecific interactions were recorded during 700 hours of observation of migrant and resident behavior in Veracruz rain forest (Rappole, field notes, 1973–1975), and the majority of those 10 were between migrant species. Powell (1980), in his observations of foraging flocks in Costa Rican cloud forest, observed 108 interactions, 3 of which were between migrants and residents. In 2 of the 3 conflicts, migrants were the aggressors. Chipley (1976), during roughly 1,000 hours of observation over a 224-day period in Colombian oak forest, observed 99 agonistic interactions. Of these, 85% were between conspecifics; only 7 involved interspecific clashes between migrants and residents, and in 4 of those 7 clashes, migrants were the attackers.

Few workers have studied interactions between resident and migrant aquatic or shrub-steppe species in the Neotropics. Myers (1980:48) did record interactions between migrant and resident shorebirds in Argentina but concluded that "complexities [of the shorebird community] make it difficult to search for any overall generalizations about migrant-resident interactions."

More work has been done on migrant-resident interactions in savannah-type habitats in Africa. Leisler (1990) performed a thorough study of the behavioral interactions between members of an ecological guild of 14 small, open-country, ground-foraging species: 7 Palearctic migrants, 5 local African residents, and 2 intra-African migrants. He found that local African residents dominated both the Palearctic and the African visitors in interspecific interactions, which were fairly frequent (101 of 502 observed interactions).

Observations of direct migrant-resident interaction are intriguing. We have speculated above that some of these interactions may occur when migrants attempt to take temporary advantage of a readily available resource on which some resident species depend (as in the case of antbirds at ant swarms). However, the circumstance that Leisler (1990) described appears to be one in which individuals of a group of migrant and resident species continue to use the same habitats and similar resources throughout a winter season, with residents uniformly dominating migrants. This situation deserves further investigation to determine details of microhabitat, food use, and survivorship of birds of the different species in the guild.

Indirect Migrant-Resident Interaction

Effects on Resident Breeding Patterns

Miller (1963), MacArthur (1972), and other authors have suggested that tropical resident species time reproduction to avoid periods when migrants are present. Keast (1980) has turned this argument around, suggesting that migrants might adjust timing of their stay in the Tropics to avoid resident breeding periods. In each case, the implication is that migrants and tropical residents as groups have the potential to affect each other significantly through competition for resources.

These predictions are logical, based on the assumption that tropical ecosystems function more or less as a steady state, that is, that resource production does not vary much through the course of an annual cycle. Under such circumstances, the influx of large numbers of species and individuals competing for resources might be expected to influence timing of breeding by the resident species—presumably to avoid competition with migrants for food resources during the critical nesting period. Many parts of the Tropics do not function as a steady state, however; they are highly seasonal. In these areas the timing of resident breeding appears to be affected more by seasonal change than by presence or absence of migrants. Skutch (1950) and Snow and Snow (1964) noted that, for the Central American avian resident community as a whole, the peak of the nesting season falls between the vernal equinox and the summer solstice, overlapping considerably with spring migration; food availability, as determined primarily by climatic factors, seemed to be the key element affecting timing of reproduction. Willis (1966:218) found that the timing of passage of migratory birds did not appear to affect the timing of breeding by antbirds on his study sites in Barro Colorado Island, Panama: The birds bred "more or less continuously during the rainy season from April to December."

This situation appeared to apply also to the resident community in rain forest of the Tuxtla Mountains of southern Veracruz. The Tuxtlas constitute the northernmost extent of rain forest in the Western Hemisphere and exhibit sharp seasonal variations in climate: A rainy season runs from June to October, followed by a season of clear periods alternating with wet weather systems from the north. By February and through the month of March, these wet systems progressively weaken, and then the brief dry season (April–May) begins. Most resident species evidently timed their breeding to coincide with the dry season and early rainy season (Andrle 1964). Though no data are available from the Tuxtlas, insect resources in many parts of the Tropics reach their peak during this period (e.g., Buskirk and Buskirk 1976; Janzen 1980). Many resident species began singing in January and February and were nesting by March or April; young were produced from May to July (Edwards and Tashian 1959; M. A. Ramos, pers. com.), roughly coinciding with the period of peak migrant passage through the Tuxtlas (late April–mid-May) (Rappole et al. 1979; Ramos 1988). Fogden (1972) made a similar observation for the resident breeding bird community in Sarawak.

Thus, there does not appear to be an overall migrant effect on resident species' breeding periods or an overall resident effect on migrant passage. However, the resident avifaunas of the Neotropical region are large and complex, and the timing of life history events remains to be determined for many resident species. For some residents, migrant passage may in fact play a key role in timing of the breeding season.

Niche Shifts

Chipley (1976), working with a migrant-resident community in oak woods of Colombia, and Rabol (1987), working in Kenyan savannah, both found that resident insectivores significantly reduced their foraging height when migrants were present. Conversely, Hutto (1980) proposed that the numbers of migrant species in different habitats in western Mexico reflect variable amounts of ecological pressure by resident species. In the habitats preferred by residents ("mature habitats"), few migrants occurred, but in those habitats avoided by residents ("disturbed areas"), large numbers of migrant species and individuals were found. Hutto (1980:198) stated, "I found that the habitats occupied by residents seem to be occupiable but avoided by migrants within the warbler guild and vice versa." He presumed that migrant warblers could have occupied niche space in mature habitats but were prevented from doing so by the "close ecological packing" of the resident warbler guild.

Two points seem important to make regarding these data. First, migrants of the warbler guild were not excluded from "mature" habitats. Five migrant species were recorded in interior, lowland, tropical, evergreen forest (versus three resident warbler guild species), and nine occurred in interior highland pine-oak woodland (versus nine residents) (Hutto 1980: Table 2). Second, the contention that a group of resident species can somehow affect habitat use by a group of migrants or vice versa is quite difficult to test on the basis of distributional information even when supplemented with data on gross insect abundance or shifts in foraging height. Too many variables are unaccounted for (Greenberg 1986; Leisler 1990).

Geographic Displacement
Moreau (1966, 1972) suggested that the winter distribution of some species of migrants might be affected by the distribution of certain tropical residents. A few other authors have presented indirect (distributional) evidence of the phenomenon (Hutto 1977), but as yet there has been no convincing demonstration of its occurrence.

ECOLOGICAL IMPACT OF MIGRANTS ON TROPICAL COMMUNITIES

Migratory birds are part of the Neotropical avifauna for several months of the year. Percentages of migrant species in tropical, terrestrial habitats vary from 4% to 50%, constituting a major percentage of some feeding guilds (Hutto 1980; Stiles 1980). Thus migrants, like other species, must affect other community members in subtle, indirect ways. They are part of the food web and energy and productivity pathways and affect the evolution of prey and predators within the communities. Is there, however, a significant "migrant effect" on tropical communities? Stiles (1980:429) stated, "Slud (1960) and others have suggested that migrants tend to complement the permanent residents ecologically, often fitting into niches seemingly unoccupied by the latter. My own observations are in full agreement—the forest canopy seems much emptier in July!" The crux of this argument is that migrants are significant components of the tropical communities that they inhabit, not a superfluous element.

As discussed above, the period of migrant absence is, in many parts of the Tropics, a period of abundance for many food categories (Skutch 1980). This abundance may be due to seasonal climatic effects rather than to movements of migrants (Skutch 1950), though migrants may play a role. An area of community dynamics in which migrants as a group may have a profound effect is in timing of fruit and seed production. Howe and De Steven (1979) found that the fruiting peak for an understory tree in tropical wet forest of Barro Colorado Island in Panama coincided with passage of transients that fed on the fruits. These transients thereby functioned as dispersal agents for the tree. The authors hypothesized that "the fruiting season of *Guarea glabra* is adaptively synchronized with northward migration of opportunistically frugivorous North American birds." Skutch (1980) found a similar relationship, as has M. A. Ramos (pers. com.: unpub. data) who observed that fruiting by *Phytolacca* (pokeweed) in Veracruz is apparently timed to coincide with the spring passage of transient Swainson's Thrushes and that few other species consume it. R. Greenberg (pers. com.) reported that fruits of *Miconia argentea* and *Didymopanax morototoni* are eaten principally by migrants, and fruits of *Lindackeria laurina* are apparently consumed almost solely by migrants in the Canal Zone (Greenberg 1981b).

Fruit ripening in some tropical plants appears to be timed to migrant wintering periods. McDiarmid et al. (1977) found that fruits of the tropical dry forest species *Stommadenia donnell-smithii* in Costa Rica were produced mostly at the end of the dry season in April. The fruits were eaten by 22 species of birds, many of which were primarily insectivorous and several of which were migrants. Morton (1971) speculated that dependence on transient bird species as dispersers may be an important strategy for many tropical forest species. Though there are many possible reasons for trees to fruit in the late dry season that have nothing to do with dispersal of fruits by migrants, the presence of migrants could have profound effects on fruit shape, size, color, nutritional content, and other aspects of the fruiting phenomenon that deserve scrutiny.

Other studies indicate comparably intricate relationships between migrants and the tropical communities they inhabit. Morton (1979) found that the Orchard Oriole was the most effective pollinator for the tree *Erythrina fusca* in a Panamanian forest. He suggested that the color and shape of *Erythrina* flowers indicate a coevolved relationship between the plant and the common, winter-resident oriole. Cruden and Toledo (1976) similarly reported that several species of Nearctic migrants were important pollinators of *Erythrina breviflora* in Michoacán, Mexico. Morton (1980) hypothesized a similar coevolved relationship between Tennessee Warblers and the flowering tropical forest vine *Combretum fructicosum,* in which the bird benefits from the plant's nectar (and perhaps, in intraspecific interactions, from the bright red "war paint" pollen that often covers the bird's face after it has fed on a blossom), and the plant benefits from the outcrossing pollination provided by the bird.

Biologists have only just begun to appreciate the importance of ephemeral resources in avian diets to the complexity of tropical communities (Levey and Stiles 1992). For example, fruit availability has far-reaching effects on avian community structure as well as on plant and insect species composition. Fruit is a resource that takes little time to consume yet can provide large amounts of energy. Omnivorous species often spend the majority of their time searching for insects, but their energy intake may be based largely on fruit consumption (Morton 1973). Without fruit (or some other ephemeral resource) to supply the energy base, these species might not be able to co-occur with insect-eating specialists because the time-energy budgets of the omnivores would, on average, run with a deficit. Conversely, insect-eating specialists may benefit indirectly by reduced competition for insects brought about by the availability of fruit, nectar, and other ephemeral resources for omnivores.

Most migrant species are not fruit specialists, but many use fruits as a substantial portion of their diet. It is not known to what extent opportunistically frugivorous migrants select highly nutritious fruit, but most of them, together with most resident species, consume small fruits that are high in carbohydrates, lipids, or both. During seasons of fruit abundance, some small migrants take smaller insects than are consumed when fruit is scarce. For example, the Bay-breasted Warbler takes significantly smaller insects during the Panamanian dry season when fruit is abundant than during the wet season. The co-occurring Chestnut-sided Warbler, which does not switch to fruit as extensively, continues to eat larger insects that satisfy its balance between energy expended hunting and energy reward per insect captured (Morton 1980).

These few examples from a subject that has received little attention from researchers indicate the degree to which migrants function as integral parts of the tropical systems in which they winter or pass through on migration. Predator-prey systems in which migrants are involved as members of the community food web have yet to be examined in detail but should yield equally fascinating coevolutionary relationships. Holmes et al. (1979) found that migrants can have important effects on larval stages of Lepidoptera during the breeding season in New Hampshire. The timing of insect hatching periods in the Tropics similarly may be related to the pressure of insectivorous migrants, causing the larval or volant stages for many insect species to be produced during periods of migrant absence.

Virtually no work has been done on the ecological impact of aquatic or shrub-steppe migrant species in Neotropical communities. Studies of this kind will be complicated in many cases by the fact that migrant and resident (breeding) populations of the same species winter in the same tropical communities.

SUMMARY

As outlined above, a hypothesis about the role of migrants in tropical avian communities was established early in the development of theoretical ecology (Willis 1966; Morse 1971; MacArthur 1972). Migrants were viewed as a satisfying paradigm of a fugitive species—nomadic ecological generalists, opportunistically skimming the excesses from tropical communities in which they could claim no membership. Even when these ideas were first proposed, however, it was recognized that they did not apply to a large number of migratory species or to all situations. Willis (1966) was well aware that Wood Thrushes were territorial in Panama and remained in tropical forest throughout the winter. Karr (1976) knew that many species of migrants were not excluded from tropical forest, including the Kentucky Warbler, with which he had done intensive work (Karr 1971b). As the number of factual studies that focus on migrants during the nonbreeding period has increased, the theoretical basis for the "migrants as interlopers in the Tropics" hypothesis, originally formulated from a competition-based understanding of community ecology, has not been sufficiently broad to encompass the variety of findings. When Greenberg (1986:281) presented his extensive review of competition in migrant birds in the nonbreeding season, he purposely eschewed the topic of migrant-resident competition, stating, "I will . . . leave the overwhelming task of analyzing resident-migrant interactions to future reviewers." Greenberg was right in recognizing that the relationship between migrants and residents is far more complex than some early models allowed. In fact, the sorting out of this relationship may amount to nothing less than solving the fundamental secrets of community structure. MacArthur (1972) and his students set us on the correct path, indicating that elemental laws govern the ways in which communities are organized, that these laws are based on natural selection, and that they can be modeled and tested. The system just happens to be considerably more complex than we had at first imagined. One final point needs to be remembered when considering the differences between migrants and residents as distinctive groupings: Not only are many species of tropical residents actually migratory to a greater or lesser extent, but nearly half of all Nearctic migrant species have resident populations that breed in the Tropics.

5
MIGRATION

NEARCTIC–NEOTROPICAL MIGRATION

Ecological studies of the journey between breeding and wintering sites are still at an early stage. Although we know most of the species that participate in these movements, at least from the Nearctic–Neotropical perspective, we still lack knowledge of when and how the journey begins, details of habitat selection and resource use at stopover sites, factors affecting timing and route selection, and the influence of sex and age on all of these aspects of the migration phenomenon.

Postbreeding Period

Of the large number of breeding bird studies that have been done, few provide information on the ecology of the species once the young fledge. In most cases, the only information available is in the form of anecdotes on a few individuals. In a justly famous account of bird migration, Cooke (1915) provided extensive documentation of dates and movement rates for a number of spring migrants. However, he devoted little space to fall migration, stating only, "Most birds care for their young until [their offspring are] old enough to care for themselves, then molt, and when the new feathers are grown start on their southward journey in their new suits of clothes" (Cooke 1915:31). Surprisingly little has been added to this account since Cooke's work. When the young fledge, do they remain on the breeding territory or move to new habitats? How long is it until they can "care for themselves"? Are there sex or age differences in postbreeding movements or habitat use? Do birds "drift southward" during the postbreeding period, as has been suggested by some authors (Hann 1937; Bent 1953; Pulich 1976) or do they remain near the breeding site until the molt is complete? Most important, when does fall migration begin, and once begun, how long does the journey take?

For some waterbirds, the details of postbreeding-premigratory movements are reasonably well understood. In the Lesser Scaup, for example, breeding pairs begin arriving on ponds in Canadian prairies by early May. Males leave incubating females by late June and generally move northward to taiga lakes, where they congregate in large numbers and molt. Females remain with young on breeding areas until the young are in the preflight stage; then adult females depart to congregate on molting lakes. The ducklings are the last to leave the breeding sites and move to join large flocks of other young at molting lakes. The considerable differences in timing of molts result in similar differences in timing of southward migrations, with drakes preceding hens, which in turn precede the year's young in migrations to wintering areas (Palmer 1975).

Massive postbreeding-premigratory movements away from breeding areas to sites where birds may remain for weeks or even months before beginning southward migration are well known from banding data on waterfowl, ardeids, and some other groups (Palmer 1962, 1975, 1976; Salomonsen 1968; Bellrose 1976; Johnson and Richardson 1982; Jehl 1990). Unfortunately, these kinds of data are not available for most Nearctic migrant passerines. Hann (1937) found that parent Ovenbirds split the brood between them after fledging, with each party following its own path, often wandering well away from the breeding territory. By the time the young reached independence (2–3 weeks postfledging), the adults had disappeared from the breeding areas, though it was several weeks before fall migration began. Pulich (1976) reported similar findings for the Golden-cheeked Warbler in central Texas. Adults split the brood at fledging but appeared to remain on the territory through the period of postfledging parental care. Young were independent at the age of 4 weeks or so, when both adults and immatures wandered into habitats not used for breeding. Migration appeared to start very early (3 July), based on records of two young birds found in northern Mexico—well south of the species' breeding range.

Nolan (1978) has provided one of the most complete accounts of postbreeding movements and behavior for a Nearctic migrant passerine, the Prairie Warbler. He found that when the young fledged at 9 to 10 days, the parents split the brood, each caring for the fledged young in their charge until the young were about 40 days old. Some females began a second nesting attempt. When this occurred, the young in the female's care joined their siblings with the male. Eighty percent of family units accompanying the male remained on or near the breeding territory during the period of postfledging parental care; 60% of the units cared for by the female remained in the vicinity. After the young achieved independence, adult males (49%) tended to remain on or near the breeding territory evidently until roughly the beginning of fall migration (September), some males (22%) moved off the territory but returned in September before migration, and a third group (29%) left the territory and either

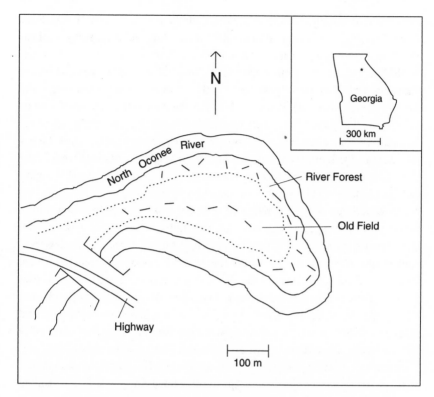

Figure 5.1. Location of river forest and old-field study sites at Horseshoe Bend study area. Dashes mark mist net locations. Adapted from Rappole and Ballard (1987).

were not seen again or did not return until the following year. Similarly, some adult females (13%) stayed on or near the breeding territory, some (6%) left and then returned before migrating in September, and the majority (69%) disappeared and were either not seen again or not seen until the following year. A fourth group of adult females (13%) either were found on the breeding territory only once or were found there and elsewhere on different occasions. Immature birds appeared to move "more or less randomly in the region for several weeks" after achieving independence from their parents (Nolan 1978:435).

A study that we performed at a site in Athens, Georgia, revealed some curious behaviors relating to postbreeding movements (Rappole and Ballard 1987). The site was contained within a loop of the North Oconee River called Horseshoe Bend (Fig. 5.1) and consists of two community types: a fringe of forest habitat along the river surrounding an island of old field habitat. From 17 April through 15 July, 33 Wood Thrushes were captured in the forest, where the species was a common breeder. However, no Wood Thrush was seen, heard, or captured in the neighboring old field habitat until 21 July. Four Wood Thrushes were captured in the field in late July and August. In contrast, no Wood Thrushes were captured in river forest from 30 July until 17 September. Individuals of seven other species (Yellow-billed Cuckoo, Ruby-throated Hummingbird, Acadian Flycatcher, Brown Thrasher, Prairie Warbler, Pine Warbler, Blue Grosbeak), known to breed in the Athens area but not found as breeders on the old field site, were captured in the field in July and August. Individuals of four other species (Northern Parula, American Redstart, Ovenbird, Scarlet Tanager), whose nearest known breeding populations were in the north Georgia mountains 100 km north of Athens, were captured showing molt and little or no fat in August in the old field habitat.

These findings suggest a range of different possible postbreeding movements, varying by species, age, sex, and even individual, most of which do not seem related to migration. This impression is further supported by observations at sites along the fall migration route on the central Gulf coast of Texas. Data from these sites indicate that the earliest passage dates for the Scarlet Tanager, Prairie Warbler, and Acadian Flycatcher are in late August, whereas the earliest dates for the Wood Thrush are in late September (Rappole and Blacklock 1985). Thus, the period between the fledging of young (early July) and actual migration (late August or September) can be quite long.

In summary, adults and young of a number of Nearctic migrant species, both passerine and nonpasserine, apparently move away from the breeding site when the young become independent or shortly thereafter. Individuals of several species move to sites in habitats different from those in which they bred, and they may remain at those sites for weeks or months throughout the molting and premigratory period. Considerably more research is needed to comprehend the details and purposes of these movements.

Transient Period

Once a Nearctic migrant leaves the breeding (or postbreeding) ground for its tropical destination, it confronts a number of situations in which natural selection likely acts heavily on birds that are unable to compete or that make incorrect choices. These situations involve timing of departure, choice of route, selection of habitats along the route for resting or for rebuilding fat reserves, and competition with conspecifics for possibly limiting resources.

Timing of Migration

All timing strategies represent compromises between reproduction and survival. The extensive work of Berthold and his colleagues with European warblers of the genus *Sylvia* has established the genetic nature of migration timing for at least some species, though the complex mixture of proximate causes affecting movements (e.g., weather) and ultimate factors affecting evolution of timing are still somewhat unclear (see review in Berthold 1993). The balances struck between reproduction and survival can be quite different for different species or even for different sex and age groups within species. As discussed earlier, most migrants do not appear to leave immediately after completion of the breeding cycle, though species that breed in the high Arctic, such as many shorebirds, are obvious exceptions. Most temperate-breeding species remain in the temperate zone at least long enough to perform the prebasic molt. Completion of molt, however, is not the sole factor governing timing of departure. Some of the factors that may affect this balance follow.

Initiation of the Breeding Cycle   The earlier a bird arrives on its breeding site, the better its chances for securing a good territory and mate and completing a breeding cycle. Rappole et al. (1979) found that several songbirds breeding in the southeastern United States were among the earliest northbound passerine migrants (Table 5.1). Individuals of these same species were also among the earliest southbound migrants. For instance, Kentucky Warblers (*Oporornis formosus*) and Louisiana Waterthrushes (*Seiurus motacilla*) begin to leave Veracruz by late March, and the first individuals of these species return to the Tropics by late July or early August.

Completion of the Breeding Cycle   Whether a bird is single-brooded or able to produce more than one complete set of offspring per reproductive cycle has important effects on when it begins fall migration. Most passerines require 60 to 70 days from commencement of nest building until the young are independent. If insufficient time remains after completion of one cycle to complete a second one, then perhaps the bird should leave as soon as possible after molting to secure the best possible wintering site (if intraspecific competition for sites is important).

Weather En Route   Deterioration of weather on the breeding site is not the only type of meteorological problem a migrant must face. Weather patterns along the route are also critical. For example, weather over the Gulf of Mexico, a major route of passage for many migrants to the Tropics, is favorable for migration only from mid-April until early October. At other times of the year, storms and unfavorable winds make trans-Gulf crossings hazardous (Buskirk 1980; Moore and Simons 1992). Timing of migration must take such factors into account. Weather can be both a proximate factor regulating immediate decisions regarding movement by the individual (Lack 1960; Nisbet and Drury 1968; Able 1973; Rappole and Warner 1976; Kerlinger and Moore 1989; Richardson 1990) and an ultimate factor serving as a selection force af-

Table 5.1

Comparison of peak migration periods for selected migrant species versus all species at study sites in two Gulf coast areas

Species	Fall Texas peak	Fall Veracruz peak	Spring Texas peak	Spring Veracruz peak	Pattern[a] Fall	Pattern[a] Spring
Empidonax flaviventris	−10	?	+20	+20	E	L
Regulus calendula	+40	?	−40	?	L	E
Polioptila caerulea	−10	?	−30	?	E	E
Catharus guttatus	+40	?	−20	?	L	E
C. mustelinus	+30	+10	−10	−20	L	E
Vireo olivaceus	−10	−10	+20	+30	E	L
Dendroica pensylvanica	?	−10	+10	+20	E	L
D. fusca	−10[b]	?	+20	+20	E	L
Protonotaria citrea	−20	−20	−20	?	E	E
Limnothlypis swainsonii	?	−10	−10	−20	E	E
Seiurus motacilla	−30	?	−20	−20	E	E
Oporornis formosus	−10	−10	−10	−10	E	E
O. philadelphia	−10	?	+20	+10	E	L
Wilsonia citrina	−10	−10	−10	−10	E	E
W. canadensis	−10	−10	+20	+20	E	L
Melospiza lincolnii	+40	?	−30	?	L	E
Zonotrichia albicollis	+40	?	−30	?	L	E

Source: Rappole et al. (1979:202).

Notes: Species were selected to illustrate specific timing patterns. Peak migration date for all species at each site is based on the 10-day period during which the maximum number of species was recorded. For the central Gulf coast of Texas, fall peak = 21–30 September and spring peak = 21–30 April. For the southern Gulf coast of Veracruz, fall peak = 1–10 October and spring peak = 11–20 April. Peaks for each species are based on that 10-day period during which the greatest number of individuals was captured, presented as the nearest 10-day period before (−) or after (+) the peak for all species at a given site. ? = insufficient data, or species does not normally occur at this site.

[a]E = early passage during migration; L = late passage.

[b]Based on small sample size (<10).

fecting overall timing of population movement (Rappole et al. 1979; Buskirk 1980; Rappole and Ramos 1994).

Relative Location of Breeding and Wintering Site At a certain level, distance can be equated with time (Cooke 1915). Birds that travel a longer distance between breeding and wintering sites should be expected to start earlier and arrive later, all other factors being equal. All other factors seldom are equal, however. Many species that breed in the same region have distinctly different timing patterns (Hagan et al. 1991). Winter latitude has been suggested as a principal determinant of timing, with species that winter at higher latitudes returning to the breeding grounds earlier, and species that winter at lower latitudes returning later (Hagan et al. 1991). This finding reflects a coarse level of analysis, which basically compares temperate migrants with tropical migrants. The evolution of migration timing is a much more complex and interesting phenomenon.

Consider the data in Table 5.1 as an example. Peak migration periods for several species of migrants are shown at two points along their migration route: southern Veracruz (latitude 18° N) and a point 800 km directly north in Texas (latitude 28° N). As Hagan et al. (1991) also found, the temperate migrants were among the earliest to depart in the spring. However, the Louisiana Waterthrush (*Seiurus motacilla*), which winters principally in southern Mexico and Central America, is also a very early spring migrant, although, in contrast to the temperate migrants, it is also a very early fall migrant. The Wood Thrush (*Catharus mustelinus*) and the Blue-gray Gnatcatcher (*Polioptila caerulea*) are both long-distance migrants to the Tropics whose breeding and wintering ranges overlap considerably. The gnatcatcher conforms to the "southeastern breeder" pattern described earlier—it is an early spring and fall migrant through Texas. In contrast, the Wood Thrush is an early spring but late fall migrant. Clearly, there is much more regarding migration timing that is worthy of study.

Intraspecific Competition for Resources at Departure, Stopover, and Wintering Sites A few species of birds are known to defend territories during migration (see the section "Stopover Site Ecology," below), and many defend winter territories (Rappole et al. 1983). This practice indicates that competition is intense for critical resources (Brown 1964). Is early arrival at stopover or wintering sites important in obtaining territories? The answer to this question is as yet unknown.

Researchers have given considerable attention to the role of competition with regard to facultative or partial migration in which one portion of the population leaves the breeding site while another does not, or one portion moves a short distance from the breeding ground while another moves a longer distance (Nice 1964; Fretwell 1972; Ketterson and Nolan 1976; Terrill 1990). In general, it is adult males that remain on or near the breeding ground while females and young move away. What is not clear yet is the degree to which these movements are mediated by direct competition for resources versus the effects on movement behavior of differential fitness components for the two classes. That is, are females and young males forced to move elsewhere by losing in competition for resources with adult males? Or do they choose to move because resources are better elsewhere, a choice that adult males must weigh against the possibility of losing their breeding territory? These questions remain a fruitful area for research (Ketterson and Nolan 1983; Gwinner 1990).

Molt The prebasic (postbreeding) molt takes 35 to 60 days for most passerines, though it can last as long as 120 days in some relatively sedentary species or populations. During this time, individuals often move to sites away from the breeding area (as discussed earlier) and to habitats different from the breeding habitat. Berthold (1988) has proposed that differences in duration of molt are related to differences in migration distance (i.e., short-distance migrants tend to have more protracted molts than long-distance migrants). Individuals of most migrant species complete the molt before initiating fall migration. Individuals of several species, however, apparently complete the body molt during migration (Winker et al. 1992b,d). Also, the adults of some species, such as some of the *Empidonax* flycatchers, molt on the wintering ground after completing their southward migration, a strategy that allows a very early departure from the breeding site. Adult Yellow-bellied Flycatchers (*Empidonax flaviventris*), which breed in boreal regions of North America, begin to appear as transients in south Texas by early August and reach their peak by early September. Immatures pass through 3 to 4 weeks later. Clearly, timing of molt can directly influence timing of migration, and vice versa.

Deteriorating Weather on the Breeding Site Cohen (1967) and many others have argued that bad winter weather is responsible for the evolution of migration for most Nearctic migrants, forcing them to evolve southward migration to increase their overwinter survival. We present an alternative theory in Chapter 6. It emphasizes a tropical origin for the evolution of migration, in which tropical species evolve a migratory habit to capitalize on available temperate resources (Rappole and Tipton 1992). Regardless of where the birds overwinter, there are obvious advantages to staying on the breeding ground as long as possible to reproduce, and to protect the breeding territory (Fretwell 1972). A few temperate residents of evident recent tropical origin illustrate this point: Carolina Wren, Northern Cardinal, and Northern Mockingbird. These birds are able to persist on their breeding territories through the temperate winter, though hard winters can cause significant die-offs. Some migrants, including many waterfowl, do not move southward until pushed by inclement weather. The majority of Nearctic migrants leave their breeding sites long before any obvious seasonal deterioration (see Table 5.1). Nevertheless, weather must serve as a key selection factor favoring continued, annual departure from the breeding site.

Ramos (1988) provides an excellent summary of the variety of timing differences that have been observed for different age and sex categories within a species. These studies indicate the importance of the effects of different life history factors on migratory movements.

The Migration Route

As is the case for the timing of migration, selection must shape the migration route (Berthold 1988). There are significant species differences with regard to the shape of the migratory route, but groups of species are known to have similar routes. Cooke (1915) recognized the basic outlines of five major routes taken by migratory species breeding in eastern North America based on ground survey data, shipboard observations and lighthouse kills:

1. *Island route*—Florida peninsula to the island chain of the Greater and Lesser Antilles to South America
2. *Bobolink route*—Florida to Cuba and across the Caribbean to South America
3. *Shorebird route*—northeast coast of North America nonstop over the western North Atlantic to South America
4. *Western Gulf route (circum-Gulf migration)*—along the Gulf coast of Texas and Mexico to Central America
5. *Gulf route (trans-Gulf migration)*—northern Gulf coast across the Gulf to Central America

Cooke believed that route 5, the trans-Gulf route, was the principal fall and spring pathway of eastern Nearctic transients that winter in Central America. He was aware that some shorebirds (e.g., Lesser Golden-Plover) followed a route north that was different from that taken south, but apparently he thought that this type of elliptical migration was exceptional and that most birds follow the same route in both directions.

Two controversies persist concerning routes of eastern Nearctic migrants. The first concerns the volume and species composition of trans-Gulf versus circum-Gulf migration (routes 4 and 5), and the second has to do with the identity of the birds following the North Atlantic route (route 3).

Table 5.2

Total net captures and relative abundance patterns of passerine migrants during fall and spring seasons at Texas and Veracruz study sites

Species	Texas		Veracruz	
	Fall	Spring	Fall	Spring
Contopus virens	77	77[a]	77	77[b]
Empidonax flaviventris	579	103[a]	59	35[a]
E. traillii/alnorum	181	14[a]	113	136[b]
E. minimus	60	5[a]	108	141[b]
Catharus fuscescens	4	31[a]	4	10[a]
C. minimus	1	74[a]	1	27[a]
C. ustulatus	27	208[a]	16	76[b]
C. mustelinus	11	79[a]	110	90
Dumetella carolinensis	26	110[a]	48	43
Vireo philadelphicus	0	9[a]	9	16[b]
Vermivora pinus	5	56[a]	3	6[b]
V. chrysoptera	0	22[a]	1	4[b]
V. peregrina	16	260[a]	10	34[a]
Dendroica pensylvanica	2	32[a]	0	13[b]
D. magnolia	12	279[a]	53	50
D. tigrina	0	1	0	0
D. caerulescens	2	2	1	0
D. virens	7	57[a]	7	15
D. fusca	2	32[a]	0	13[b]
D. castanea	2	62[a]	2	30[a]
D. striata	1	0	0	0
Mniotilta varia	131	207[a]	75	18[a]
Setophaga ruticilla	21	71[a]	14	24
Helmitheros vermivorus	2	118[a]	30	23
Limnothlypis swainsonii	4	48[a]	11	6[b]
Seiurus aurocapillus	38	515[a]	92	77
S. noveboracensis	41	184[a]	21	19
S. motacilla	6	24[a]	24	2[a]
Oporornis formosus	19	340[a]	272	123[a]
O. philadelphia	147	54[a]	2	44[a]
Geothlypis trichas	12	232[a]	43	98[b]
Wilsonia citrina	6	171[a]	216	105[a]
W. pusilla	208	21[a]	99	48[a]
W. canadensis	245	476[a]	69	69

Source: Rappole et al. (1979).

Note: Differences between fall and spring samples for each study were tested using a chi-square analysis for equality. Expected values used in the test were obtained by summing fall and spring samples for a given area and dividing by 2. Samples were corrected for net-hour bias using total net-hours per season for each area (see text). Except where noted, numbers are based on birds netted.

[a]Observed values are significantly different from expected ($p < .05$).

[b]Sample is biased because of selective collecting. No statistical analysis was performed.

Trans-Gulf versus Circum-Gulf Migration   The controversy over the species composition of trans-Gulf versus circum-Gulf migrants began in the 1940s with published interchanges between George Lowery (1945, 1946, 1951, 1955) and George G. Williams (1945, 1947, 1950, 1951). Lowery asserted that most species from eastern North America were trans-Gulf migrants in fall and spring, as Cooke (1915) had proposed. Williams maintained that observations along the Texas Gulf coast documented that many of these species followed a circum-Gulf route rather than a trans-Gulf one. Work by Stevenson (1957) appeared to resolve the dispute, mainly in favor of Lowery's ideas. As transient distributional data have continued to accumulate, however, this question does not appear to have been completely resolved. Intensive studies along the western Gulf coast in Texas and Mexico have shown marked seasonal differences in volume of migration for members of the same species (Table 5.2). Several species show differences of an order of magnitude or more in the amount of fall versus spring migration. For instance, 2 Gray-cheeked Thrushes (*Catharus minimus*) were captured during fall migration in 60,000 net-hours in Texas and 28,040 net-hours in Veracruz. Spring migration values were strikingly different: 74 were captured in 35,100 net-hours in Texas and 27 in 27,670 net-hours in Veracruz. These and similar data indicate that migration routes are probably elliptical for a large number of transients from eastern North America (Rappole et al. 1979; Winker et al. 1992d; Rappole and Ramos 1994), with an easterly trans-Gulf route in fall and a westerly circum-Gulf route in spring (Fig. 5.2).

Migration over the Western North Atlantic   The North Atlantic, or shorebird, route (route 3) has been recognized as an important fall pathway for shorebirds migrating to South America (Cooke 1915; Wetmore 1926; Richardson 1976). However, shipboard observations (Scholander 1955; Drury and Keith 1962; McClintock et al. 1978), sightings in Bermuda (Nisbet et al. 1963; Wingate 1973), radar studies (Drury and Keith 1962; Williams and Williams 1978; Larkin et al. 1979), and physiological analyses (Nisbet et al. 1963) indicate that some passerines may also follow this route to wintering grounds in the Caribbean and South America. In particular the Blackpoll Warbler (*Dendroica striata*) is thought to use this route (Fig. 5.3). Murray (1989) has contested this conclusion, contending on the basis of distributional and fat data that blackpolls follow a coast-hugging route from extreme northeastern Canada to the southeastern United States, thence across the water to South America. Murray (1976) has proposed that birds found offshore are disoriented. Unfortunately, distributional and fat data cannot resolve the question. Nisbet's data (Nisbet et al. 1963; Nisbet 1970) demonstrate that the crossing is possible and that some portion of the population probably does it. Observa-

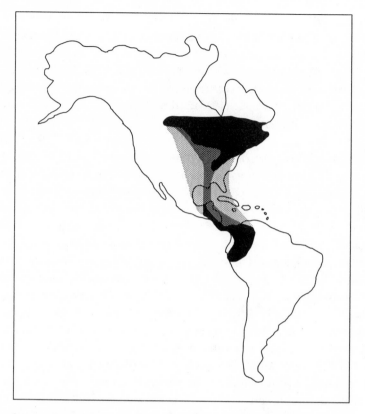

Figure 5.2. Hypothetical migration route of the Blackburnian Warbler (*Dendroica fusca*). The easterly trans-Gulf route in fall and the more westerly circum-Gulf route in spring are shown as gray shading, the overlap of the two routes represented by the darker gray shading. Breeding and winter ranges for the species are shown in black. Adapted from Rappole (1991).

tions by Murray (1965, 1966, 1989) point out that many blackpolls do not follow the western North Atlantic route. Radio-tracking data will likely be necessary to settle the issue. In any case, the North Atlantic route is certainly elliptical, regardless of the species that follow it, because the winds, which are often favorable in fall, are not favorable for such a crossing in spring.

Trans-Atlantic Migration There is another route for Nearctic migrants from eastern North America of which little is known. Some individuals of two species of Nearctic migrants (Red-eyed Vireo and Summer Tanager) have been found to winter in the equatorial forests of Gabon, West Africa (Brosset 1968). The Northern Wheatear (*Oenanthe oenanthe*) performs a similar mi-

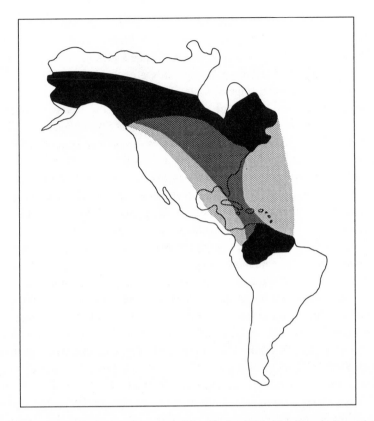

Figure 5.3. Hypothetical migration route of the Blackpoll Warbler (*Dendroica striata*). The western North Atlantic route in fall and the trans-Gulf route in spring are shown as gray shading, the overlap of the two routes represented by the darker gray shading. Breeding and winter ranges for the species are shown in black. Adapted from Rappole (1991).

gration, breeding in Greenland and wintering in Africa. These trans-Atlantic migrations deserve further study.

Age Differences in Migration Route   Just as there are differences among age and sex groups of the same species in terms of migration timing, there are also differences among these groups in choice of migration route. In particular, there are marked differences between adults and immatures of the same species. These differences are most clearly represented in the relative proportions of young and adults at inland and coastal sites. Based on data from television tower kills and intensive netting studies, inland sites tend to have adult : immature ratios between 1:1 and 1:2 (Ralph 1978; Winker et al. 1992d), whereas coastal sites often have significantly higher proportions of immatures (Robbins et al. 1959; Drury and Keith 1962; Murray 1966; Stewart et al. 1974). Ralph (1978:237) has referred to this phenomenon as the "coastal effect" and suggests that "these young birds are disoriented and thus have strayed from the normal [inland] route of the species." Rappole et al. (1979) observed the "coastal effect" for some species along the Texas coast but not for others. They proposed an alternative explanation: "We hypothesize that the disproportionately large number of immatures captured is the result of the inability of immatures to find and compete as effectively for the resources necessary for fattening preparatory to a long over-water flight. These immatures choose a longer circum-Gulf route where resources are available rather than the trans-Gulf route where resources are not available" (Rappole et al. 1979: 210). As of yet, these different hypotheses have not been thoroughly tested to determine which is more likely to reflect reality or whether other, completely different explanations are more valid.

Indirect Migration Routes   Individuals of some species appear to follow routes that do not represent the shortest distance between two points. The Blackpoll Warbler, for instance, appears to head eastward on its fall migration across Canada from the western portions of its range before turning southward (Nisbet et al. 1963). Cooke (1915) provided additional examples (Red-eyed Vireo, Western Tanager, Marbled Godwit, Cliff Swallow). He gave two separate classes of explanations for indirect migration routes, one or the other of which have been used to explain the shape of most migration routes: (1) Indirect routes are the result of historical factors, such as recent range expansion or climatic change, on which natural selection has not yet had a chance to operate (Lowery 1945; Wolfson 1948), and (2) indirect routes are favored over direct routes because of various selection factors, such as avoidance of overwater flights, deserts, or persistent cold spring weather in the high plains and mountains of western North America (Berthold 1988). As more data accumulate on the ecology of migration, it seems probable that the historical explanations will be replaced by ecological ones.

In some instances, historical and ecological explanations may overlap. For instance, Arctic Warbler (*Phylloscopus borealis*) populations that breed in Alaska migrate to the Philippines, a distance of 7,500 km, whereas a flight of 6,800 km would take them to the Neotropics. A historical explanation seems logical here, assuming that the Arctic Warblers of western Alaska are derived from Siberian birds and are following the ancestral (genetically programmed?) migration route. However, it could also be argued, on the basis of a tropical origin of the Arctic Warbler in Southeast Asia, that the birds are returning to the part of the Tropics where they evolved and where they can fit into an existing tropical community. If they were to migrate to the Neotropics, presumably they would have to survive as wanderers, subsisting on temporarily superabundant resources.

Factors Affecting Evolution of Route Shape   The principal factors affecting route shape appear to be as follows:

1. Relative locations of breeding and wintering areas. All else being equal, the route should be the shortest distance between the two endpoints.
2. Avoidance of flights over areas lacking in suitable stopover sites—areas such as large bodies of water for landbirds, mountains for lowland species, deserts for forest species.
3. Preference for pathways that provide dependable tailwinds or other aids to migratory movement, such as thermals for raptors.
4. Avoidance of pathways in which headwinds or storms are likely to occur.

We propose that the shapes of most migration routes reflect compromises among these factors.

Stopover Site Ecology

The study of stopover site ecology is in its infancy. Though hundreds of thousands of transients have been banded, and a number of publications have been produced using these data, the majority of these publications deal with the gross outlines of the migration phenomenon—namely, which birds are going where and under what meteorological conditions. Navigation, orientation, and the physiology of migration have been thoroughly studied (Emlen 1975; Wiltschko and Able 1988), but investigation of how they relate to what an individual bird is doing has lagged behind. This state of affairs is not surprising, given the daunting technical problems of examining key ecological parameters for individual transients (Terrill 1988). Nevertheless, we cannot claim to understand the biology of even one species until this portion of the life cycle has been illuminated.

Individuals of most species of transients migrate at night unless crossing some vast expanse of water or desert, evidently to avoid the air turbulence caused as the sun heats the earth's atmosphere and perhaps to minimize the threat of predation. Exceptions include ducks, ardeids, hawks, swallows, and a number of short-distance or facultative migrants. When nocturnal migrants stop to rest for the day, how selective are they with regard to stopover habitat? This question is vital to an understanding of migration because wrapped up in the word "habitat" is a whole range of potentially critical resources, including food abundance, presence of competitors, vulnerability to predators, and shelter from inclement weather (Moore and Simons 1992).

The obvious answer to the question of migrant habitat preference would be that, given a choice of two habitats in which to spend the day, one of which provides suitable foraging microhabitats and the other of which does not, the bird will choose the habitat with appropriate foraging microhabitats. However, many studies, as well as the experience of most birdwatchers familiar with migration, have shown that transients often do not appear to be particularly selective about stopover sites and that migrants occur in a variety of unlikely places (Rappole and Warner 1976). Power (1971:442) offered this explanation of the phenomenon: "Birds in migration may tend to replenish energy stores as quickly as possible by foraging in zones where food is most abundant, without engaging in behavior that minimizes overlap with other species."

Other studies have demonstrated that transients, at least in some instances, are selective about stopover habitats. Parnell (1969) found that several species of wood warblers demonstrated "some degree of habitat selection" during their stopover at his North Carolina study sites. Berthold and his colleagues obtained similar results in their examination of habitat selection by transient passerines (Berthold et al. 1976; Bairlein 1983, 1985, 1988, 1990, 1992; Berthold 1988). They found that transients had distinct species-specific distribution patterns by habitat at their stopover sites. Winker and his co-workers in Minnesota found that not only did transient species demonstrate habitat preferences but these preferences also could change between seasons at the same site (Winker et al. 1992b–d), an occurrence also reported by Jones (1985) and Hutto (1985b). Moore and his associates have studied migrants along the northern coast of the Gulf of Mexico in Mississippi, where they have found evidence of preferences for different habitat types by different species (Moore and Simons 1992).

None of these studies has taken into account the possible effect of the transient's physiological state on its choice of habitat. The study of the physiological states of migrants has been quite thorough—at least in the laboratory (King 1961; King and Farner 1963; King 1972). Groebbels (1928) proposed the terms "Zugstimmung" and "Zugdisposition" to refer to two distinct physiological states experienced by transients during the course of a long migratory journey. Birds in Zugdisposition are hyperphagic, increasing food intake by as much as 40% and storing the excess energy as subcutaneous fat reserves. Birds in Zugstimmung are ready to begin and sustain a migratory flight. Zugunruhe, or migratory restlessness, is the behavior of a migrant held in captivity when it is in Zugstimmung. Zugunruhe has often been used in the study of navigation and orientation (Emlen 1975). In addition to Zugstimmung and Zugdisposition, Jenni and Jenni-Eiermann (1992) have proposed a third physiological state for transients, a fasting state during stopover. These researchers have demonstrated that plasma levels of key metabolites are distinctive for all three states.

There has been some disagreement about when birds enter Zugdisposition, with some researchers arguing that birds are likely to become hyperphagic at or near their breeding or wintering site (Rappole and Warner 1980), while others have argued that the birds move away from these sites to special feeding areas, or perhaps even begin northward or southward movement before laying down fat reserves (Odum et al. 1961; McNeil and Carrera de Itriago 1968). In any case, as Berthold (1975:99) noted in his review, "Each migratory period, as a rule, is preceded by or associated with the development of a specific physiological state leading finally to a general metabolic condition in which there is sufficient energy available for migration."

For most long-distance migrants, the amount of reserves stored during premigratory Zugdisposition is insufficient to allow them to make the entire flight without rebuilding reserves. The maximum nonstop flight distance that has been proposed for a Nearctic passerine is the 80-hour, 4,000-km journey of the Blackpoll Warbler (Nisbet et al. 1963). Obviously, individuals of most species make much shorter hops between refueling stops than that. Thus, migration must involve "a wavelike alteration of migrating and feeding activity" (Berthold 1975:98) for most transients. Are the habitat needs for birds in Zugstimmung (flyers) the same as for those in Zugdisposition (feeders)?

We performed a study in Texas along the western coast of the Gulf of Mexico in which we attempted to examine the behavior of transients at a stopover site where they were known to neither breed nor winter (Rappole and Warner 1976). Fifty mist nets were placed along an 11.4-ha strip of river forest dominated by hackberry (*Celtis laevigata*) and cedar elm (*Ulmus crassifolia*) bordering the Aransas River, 48 km north of Corpus Christi and 10 km inland from the Gulf of Mexico. We netted for two fall and two spring seasons, August–October and March–May 1973–1975, for a total of 95,100 net-hours. We found, as have many other researchers, that fall migration proceeded as a series of sharp peaks in species abundance associated with fronts, interspersed with periods of relatively low abundance. For instance, a total of 415 Yellow-bellied Flycatchers were captured from 15 August to 21 October in 1973, but 114 of them were captured on a single day, 15 September (Winker and Rappole 1992; Rappole, unpub. data).

We observed a striking contrast in the behavior of birds at our site during the peaks as opposed to the times of low abundance. Observations of Yellow-bellied Flycatchers and several other species during peaks indicated that the birds arrived in flocks of hundreds of individuals. Most individuals appeared to continue their migratory flight at sunset, and within a day or two after their arrival, the woods was emptied of these flocks until the occurrence of the next peak. The behavior of birds in these flocks contrasted distinctly with the behavior of conspecifics that remained on the site in the days following the departure of the majority of the birds. Birds occurring in flocks during peaks were active and gregarious, were not particularly selective with regard to habitat, and remained in an area for only a short period. Birds remaining after the departure of the flocks tended to be solitary, aggressive, and selective with regard to habitat, and they remained on a given site for several days, even on occasion returning to a specific net if captured and moved to another location.

We examined this contrast more closely for one species occurring as a transient at the site, the Northern Waterthrush. We found that, during peaks, Northern Waterthrushes were captured in nets throughout the dry upland woods, though most such habitat was poor at best for the species. In periods between peaks, waterthrush activity was focused around a 0.6-ha temporary pond in the middle of the woods. In addition to the 50 nets placed elsewhere in the woods, during the spring of 1975 (10 April–16 May) we placed nets around the border of the pond. Each bird was weighed at capture, then recaptured and reweighed each morning for the duration of its stay, until all waterthrushes had departed the woods. Of the 122 Northern Waterthrushes captured in the woods, 20 were recaptured in the 10 nets at the pond, and only 6 were recaptured in the 50 nets away from the pond. Simultaneously, we observed the behavior of the birds from an elevated blind in the center of the pond throughout the day. We found that transient Northern Waterthrushes were establishing territories on the pond border. These territories were defended using the *chink* call note, chases, and distinctive stylized displays, as Schwartz (1964) described in his observations of territorial encounters in transient and wintering waterthrushes in Venezuela. Seventeen marked birds held territories during the course of the study. Most birds were not successful in obtaining a territory on arrival—average length of stay was 5.6 days, and average length of time holding a territory was 3.2 days. Known territory holders showed a mean weight gain of 0.68 g, whereas other recaptured individuals showed mean weight losses of 0.28 g. Maximum weight gain in a 24-hour period for a territory holder was 0.9 g (7% of body weight). Similarly, recent studies have shown that, on average, the individuals of some species gain weight over the course of a day in a given stopover habitat while those of other species show little or no weight gain (Moore and Simons 1992; Winker et al. 1992b).

Based on these data, we proposed that transients occur at stopover sites in at least two different physiological states, Zugstimmung (flyers) and Zugdisposition (feeders). We hypothesized that behavior of the individuals in these different states is distinctly different, even for individuals of the same species. Small, arthropod-eating passerines in Zugstimmung (a flying state) may be active, gregarious, and relatively nonselective with regard to habitat; may show no site fidelity; and may remain on a given site for only one or two days. We predicted that these birds are likely to show average weight losses during their stay. In contrast, individuals in Zugdisposition (a feeding state) may be solitary, aggressive toward conspecifics (territorial in some cases), and highly selective with regard to habitat and may tend to show average weight gains during their stay. We hypothesized that each individual transient alternates between Zugdisposition and Zugstimmung throughout the journey between breeding and wintering sites in a complex interplay of individual experience, competitive ability, habitat availability, resource availability, and weather (Fig. 5.4). Testing of this hypothesis will require a simultaneous field assessment of the physiological state of the individual with its behaviors (e.g., habitat selection and territoriality), a daunting task at present. Several predictions, however, can be made based on the hypothesis, and they can be tested.

Hypothesis: Transients alternate between the physiological states of Zugdisposition and Zugstimmung during the course of a long-distance migration.

Prediction 1: The majority of migrants at a particular stopover site will be relatively nonselective with regard to habitat.

Prediction 2: The majority of migrants at a stopover site will remain for less than 12 hours, weather permitting, and will show little or no net weight gain.

Prediction 3: A small (variable) percentage of migrants will stay for longer periods at a stopover site.

Prediction 4: The individual migrants that stay for longer periods at a stopover site will show significant net weight gains and will be highly selective with regard to habitat.

The findings presented above demonstrate that transients exhibit a broad range of habitat selection behavior—from apparently not selective to highly selective. These findings suggest four possible explanations for the observed range in stopover behavior:

1. Preferred habitat is not available at termination of the night's flight, so the birds set down in what is available.

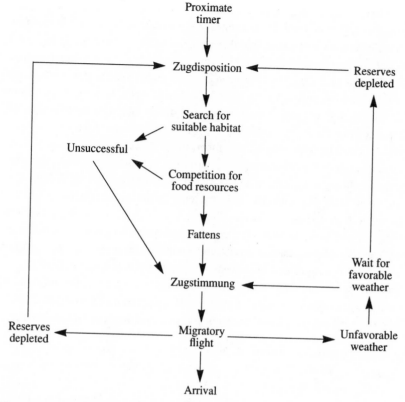

Figure 5.4. Hypothetical sequence of events for an individual migrant moving to or from its wintering site.

2. Migrants are such generalists that they can exploit whatever habitats they land in.

3. Individuals arrange themselves in the array of available habitats according to the Fretwell-Lucas-Brown model of habitat selection, filling the most suitable habitats first, based on the presence of conspecific competitors.

4. Any group of transients occurring at a site represents a mixture of individuals in different physiological states. Those in Zugdisposition (feeding state) are likely to be highly selective with regard to habitat. Those in Zugstimmung (migratory state) may not be particular about habitat choice, because food may be relatively unimportant (as compared with other factors, such as avoidance of competition with conspecifics, unnecessary expenditure of energy in low-return foraging behavior, or predator avoidance).

None of these explanations excludes the others. All of these factors could play a role during any given migratory journey. However, at present no data exist to document the relative importance of each.

We tend to assume that the phrase "preferred habitat" for transients encompasses all of the critical aspects for a species (food, protection, shelter, and so forth), but this assumption is not necessarily true. Hutto (1985a) and Moore and Simons (1992) have provided an illuminating discussion of the factors that can affect migrant habitat selection. They point out that some factors that can affect selection are "intrinsic" to a habitat (e.g., food abundance, habitat structure, habitat patchiness, intraspecific competition, interspecific competition, and predation), and some factors are "extrinsic" to the habitat (e.g., habitat availability, weather, and time available to the individual transient). The authors also make the case that a bird's energetic state (i.e., fat reserves) can affect habitat choice. This last point is important. Bergman (1994), for instance, found in a radio-tracking study that nonbreeding populations of Northern Pintails foraged on flooded stubble fields along the central Texas coast but rested in open bay waters of the Laguna Madre, 3 to 5 km away, presumably to avoid predators. These findings imply that a particular habitat's suitability may depend to some extent upon the specific needs of the individual at a given moment. That is, a bird needing to rebuild fat reserves may prefer one habitat, whereas another of the same species needing a safe, sheltered place to wait for the next night's flight may prefer a different habitat (Lindstrom 1990).

SUBTROPICAL – NEOTROPICAL MIGRATION

The list of 338 species that we have included as Nearctic migrants incorporates few species that migrate from the subtropics into the tropical zone (see Appendix 1), even though such migrants would meet our definition of having breeding areas north of the tropic of Cancer and wintering areas south of that line. Findings from the historical literature and some recent studies raise questions as to whether a large segment of this avifauna may not, in fact, qualify as Nearctic migrants. Our list does include some of the obvious ones already well recognized as migrants in the literature—for example, Sulphur-bellied Flycatcher—as well as the West Indian migrants (White-crowned Pigeon, Gray Kingbird, Black-whiskered Vireo, and the like). The reason these particular species are obvious migrants is that they disappear completely from their subtropical and tropical breeding sites in Texas, northern Mexico, and the West Indies. Other species may not disappear completely, or they may leave only certain habitats or only in certain years. For instance, Vega and Rappole (1994) studied an avian community in south Texas in which a large percentage of the birds reported in the literature to be "permanent residents" in south Texas (Rappole and Blacklock 1985) disappeared from the study site from August to April. Where these birds went is unknown, but many probably migrated south, perhaps to wintering quarters in the Tropics (where resident populations of several of these species occur). Without extensive banding-recovery studies or actual observations of birds in transit (e.g., lighthouse or television tower kills), the migrations of these birds could go unrecognized, particularly because most of the smaller species would probably be nocturnal migrants.

Wetmore (1943) and Thiollay (1977, 1979) observed individuals of a few subtropical- and tropical-breeding species, most of which were large diurnal birds, in apparent migration along the Veracruz coast. We present a list of these probable subtropical–Neotropical migrants in Table 5.3, a list that is almost certainly incomplete. It is of interest to note that one of the species listed in Table 5.3, the Snail Kite, has been reported as having a complex austral migration pattern. Hayes et al. (1994) suspect that this species is represented in Paraguay by breeding birds that winter north of Paraguay and wintering birds that breed south of that country.

INTRATROPICAL MIGRATION

As in the case of subtropical–tropical migration, intratropical migration has been acknowledged for a few species for some time. Yellow-green Vireos, for instance, have been recognized as intratropical migrants, breeding in Middle America from March through August, then migrating to wintering grounds in South America (Morton 1977). Recent studies have documented that many other tropical "resident" species disappear from areas, habitats, or regions for entire seasons. Winker et al. (1992a) documented apparent movement of "resident" Lesser Yellow-headed Vultures (*Cathartes burrovianus*) in southern Veracruz. Levey and Stiles (1992) summarized much of the literature on this phenomenon, noting that migratory movements are much more common among Neotropical residents than has been recognized, an observation borne out by several studies (e.g., Slud 1964; Karr 1976; Ramos 1983, 1988; Stiles 1985a,b, 1988; Loiselle and Blake 1991). Stiles (1983) proposed that more than half the avifauna of Costa Rica undergoes seasonal movements.

Some of these movements have been found to occur over relatively short distances. Ramos (1983) reported evident altitudinal migration, apparently related to weather, for some highland species in Veracruz. Several individuals of the following species were captured at lowland study sites (<150 m elevation),

Table 5.3

Species that breed in subtropical northern Mexico with probable migratory populations that winter in the Neotropics and whose movements are unrecognized by the 1983 AOU *Check-list*

Species	Common name	Type of evidence (and source)[a]
Tachybaptus dominicus	Least Grebe	Flocks on lakes in spring (1)
Phalacrocorax olivaceus	Olivaceous Cormorant	Fall and spring coastal flights (3,4)
Eudocimus albus	White Ibis	Fall coastal flights (3)
Ajaia ajaja	Roseate Spoonbill	Fall coastal flights (3)
Mycteria americana	Wood Stork	Fall and spring coastal flights (1,3–5)
Coragyps atratus	Black Vulture	Fall and spring coastal flights (3,4)
Rostrhamus sociabilis	Snail Kite	Spring coastal flights; fat birds collected along coast in spring (4,5)
Buteogallus urubitinga	Great Black-Hawk	Fall coastal flights (3)
Buteo brachyurus	Short-tailed Hawk	Fall coastal flights (2)
B. albicaudatus	White-tailed Hawk	Spring coastal flights (5)
Columba flavirostris	Red-billed Pigeon	Fall coastal flights (3)
Nyctidromus albicollis	Common Pauraque	Subspecific differences in specimens (5)
Tyrannus couchii	Couch's Kingbird	Spring coastal specimen (5)
Molothrus aeneus	Bronzed Cowbird	Spring coastal flights (5)

Source: Based on Rappole et al. (1994).

Note: AOU *Check-list* = American Ornithologists' Union (1983).

[a]Numbers in parentheses refer to the following citations: (1) Andrle 1967; (2) Andrle 1968; (3) Thiollay 1977; (4) Thiollay 1979; (5) Wetmore 1943.

though they are reported to be highland species (>500 m) (Loetscher 1941): White-throated Robin (*Turdus assimilis*), Slate-colored Solitaire (*Myadestes unicolor*), Blue-crowned Chlorophonia (*Chlorophonia occipitalis*), Common Bush-Tanager (*Chlorospingus ophthalmicus*), and Slate-throated Redstart (*Myioborus miniatus*). Captures generally corresponded to the occurrence of cold fronts. Similar altitudinal migrations have been documented for Costa Rica by Stiles (1988), Loiselle and Blake (1991), and others (see review in Levey and Stiles 1992).

The studies by Powell and his colleagues provide remarkable documentation of seasonal movement in the Resplendent Quetzal (*Pharomachrus mocinno*) in the highlands of Costa Rica, Mexico, and Guatemala (Jukofsky 1993; Powell and Bjork 1994). Using radio tracking, Powell and his associates have been able to follow quetzals through an annual cycle. In Costa Rica the birds breed in lower montane rain forest (1,350–1,550 m elevation) (Holdridge 1987) of the Monteverde Cloud Forest Preserve from February through June. In July the birds move 1 to 4 km westward into fragments of premontane moist forest at 1,300 to 1,450 m elevation. They remain on these sites until October, when they move 8 to 30 km north and east into premontane wet forests at 700 to 1,200 m elevation. They stay on these sites until January, when they return to their breeding area in cloud forest of the Monteverde Preserve (Fig. 5.5). Quetzals evidently migrate to track the fruiting periods of Lauraceae trees, their principal food source (Skutch 1944; Wheelwright 1983; Wheelwright et al. 1984).

The discoveries of Beebe (1947) in a mountain pass in Venezuela and McClure (1974) in mountain passes in Malaysia and elsewhere in Southeast Asia indicate that Powell's findings regarding quetzal migration are not unique. Both Beebe and McClure reported extensive movements of lowland tropical resident species at night in apparent migratory flight. McClure (1974) captured individuals of 47 tropical endemic species using mist nets and bright lights at Dalton Pass in the Philippines (1,400 m elevation). His account of the Hooded Pitta (*Pitta sordida*) is typical: "A forest pitta widespread in SE. Asia from the Himalayas through Indochina, Philippines, Malaysia, Indonesia into New Guinea. It is rarely seen, and secretive like other pittas. Thousands cross the mountains of the Philippines and Malaya but we know little more about them" (McClure 1974:192). These studies confirm the existence of intratropical migration, though they leave the questions of origin, destination, and purpose unanswered.

Levey and Stiles (1992) have hypothesized that most intratropical migrants are frugivores or nectarivores, like the quetzal, that are forced to move by seasonal variation in fruit or nectar resources. In fact, the authors propose that Nearctic migrants to the Tropics evolved from frugivorous and/or nectarivo-

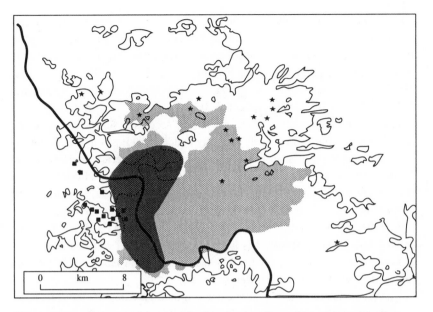

Figure 5.5. Migratory movements of the Resplendent Quetzal in northwestern Costa Rica in the vicinity of the Monteverde Cloud Forest Reserve (light shaded area). Cloud Forest breeding areas (January–July) are indicated by dark shading; squares show bird locations in premontane moist forest (July–October) across the Continental Divide (heavy line); stars show bird locations in premontane rainforest (October–January). Thin lines enclose the forested regions. Adapted from Powell and Bjork (1994).

rous ancestors (see Chapter 6). Is there evidence that species exploiting other types of resources also undergo regular, seasonal, intratropical movements? This question is a difficult one to answer unequivocally, not because there are no insectivores known to be intratropical migrants but because few migrant species can be shown to restrict their diets to one resource type. Many birds, including a number of Nearctic migrants (Morton 1980), exploit arthropods during one season or at one site but use fruits or nectar in another season or site. Sorting out the importance of the different kinds of foods that intratropical and intercontinental migrants use remains an important area for research.

AUSTRAL MIGRATION

Chesser (1994) defines austral migrants as "those species that breed in the southern parts of South America during the austral summer, and migrate north, towards or into Amazonia, and in a few cases over Amazonia, to spend the southern winter." Reports of austral migration in South America date from the early 1800s (Azara 1802–1805) and cover most regions and countries of the continent (Barrows 1883; Hudson 1920; Zimmer 1931–1945, 1938, 1947–1955; Olrog 1963b; Johnson and Goodall 1967; Sick 1968; Capurro and Bucher 1988; Willis 1988; Hayes et al. 1994).

Phelps and Phelps (1958, 1963) reported 10 species or subspecies as austral migrants wintering in Venezuela, Ridgely and Tudor (1989) listed 25 species of oscine passerines as austral migrants in South America, and Hilty and Brown (1986) cataloged 39 species for Colombia. Similar allusions to the phenomenon are given by Johnson and Goodall (1965, 1967), Haverschmidt (1968), Gore and Gepp (1978), Meyer de Schauensee and Phelps (1978), Parker et al. (1982), Belton (1984, 1985), Sick (1984), and Fjeldså and Krabbe (1990). In addition, a few studies have reported on the movements of individual austral migrants (Lanyon 1978, 1982; Remsen and Hunn 1979; Remsen and Parker 1990; Marantz and Remsen 1991).

The phenomenon is much more extensive than these reports indicate, however. Hayes et al. (1994) reported 80 austral migrants for Paraguay alone. Chesser (1994), in his thorough review of the subject, found evidence of austral migration in 220 South American species and believed that this number was probably quite conservative.

Chesser (1994) stated that the austral migration system has two significant and perhaps unique characteristics: (1) Roughly 75% of the species involved in austral migration have populations that undertake partial, nondisjunct migration, and (2) many species that have austral migrant populations or subspecies also have tropical resident populations or subspecies. As we learn more about the migrations of birds from the northern subtropics into the Neotropics, we may find that these apparent distinctions between the northern and southern migration systems of the Western Hemisphere are less marked than they appear to be at present.

Most data documenting austral movements derive from intensive observations. This situation is quite different from that in North America, where most data derive from extensive banding operations. No continent-wide banding system or centralized repository for banding data has yet been developed for South America. Nevertheless, some significant individual banding efforts have been made, particularly those of Olrog (1963a, 1968, 1969, 1975), who was able to plot the migratory pathways for several waterbird species from northern Argentina into southern Brazil using recapture data. Brazil has also initiated a government banding program, which should yield important insights into austral and intratropical movements.

SUMMARY

The migrant life cycle can be divided into the following segments: breeding, postbreeding-premigratory, fall transient, winter resident, and spring transient. Of these phases, only the breeding period is reasonably well understood. The postbreeding-premigratory phase, extending from the time the last brood fledges until the birds leave on southward migration, is poorly studied. The small amount of knowledge available on the topic indicates that behavior, movements, and habitat use during this time can vary by age, sex, and species and perhaps even from individual to individual. These factors also affect the timing of a bird's departure southward. Major differences in timing of migration among species appear to be genetically programmed, based on selection-forced balances between reproductive success (favored by early arrival on and late departure from the breeding ground) and survival (favored by early departure from and late arrival on the breeding ground). Route selection appears to be a similarly programmed phenomenon for many long-distance migrants. The pathways followed by different species can be quite dissimilar, even when their breeding and wintering ranges are alike. Natural selection appears to shape the routes through the effects of prevailing winds, storm probabilities, physical barriers, and occurrence of specific habitats en route. The need for specific habitats as stopover sites for refueling during migration has long been recognized for waterfowl and shorebirds but is less understood for songbirds. Considerable work remains to be done to determine the importance of stopover sites from both ecological and conservation perspectives. The occurrence of subtropical–tropical, intratropical, and austral migration has been obscured by the persistence of portions of populations on the breeding range and by a general lack of banding studies. Nevertheless, evidence is increasing that seasonal movements are common in these avifaunas.

6

MIGRANT EVOLUTION

Migrants are a collection of paradoxes. They often leave their summer homes long before food is scarce there (Cooke 1915; Rappole et al. 1979; Gauthreaux 1982). Then they travel thousands of kilometers in uncertain weather over hostile environments, stopping at intervals to rebuild fat reserves. Finally they arrive at tropical destinations whose habitats are often radically different from those left behind in the temperate zone. There, migrants enter complex, species-rich tropical communities where they must compete for resources with highly specialized endemics, often at periods of the year when resources are at a low ebb (Hespenheide 1975; Emlen 1977; Morton 1980; Orejuela et al. 1980; Stiles 1980; Waide 1980). After accomplishing this feat and surviving for six months in these communities, they head north, returning to their temperate breeding areas at a time when resources in many parts of the Tropics are approaching their peak (Skutch 1950, 1980; Snow and Snow 1964). The solutions that have been put forward to explain this biological puzzle are numerous. We present and discuss several below.

THEORIES ON THE ORIGIN OF MIGRATION

Past discussions of the origins of migration have generally focused on two opposing concepts: (1) a northern ancestral home theory, in which the ancestors of migrants originally spent the entire life cycle in what is now the temperate zone, and (2) a southern ancestral home theory, in which migrant ancestors originated in the Tropics (Cooke 1915). However, Gauthreaux (1982) has noted that a diverse mixture of explanations has been grouped under these two categories. Thus, consideration of the evolution of migration based on point of origin may not be as helpful as some other perspectives.

The one factor that all of these theories have in common is an ancestral sedentary population. As Gauthreaux (1982:92) pointed out, "An animal should not move if the location where it was produced remains suitable." All of the theories begin with something that causes these ancestors to move. It is

this something that constitutes the key difference among the theories address-
ing the origin of migration, and therefore we use these "motive forces" as a
means of categorizing the various theories.

In his thorough review of the topic, Gauthreaux (1982) stated that the most
important advance in consideration of theories on the origin of migration was
the recognition that natural selection shapes the appearance of migration
through the optimization of reproduction and survival (Lack 1954). We inter-
pret this statement to mean that the current breeding and winter ranges, routes,
timing, and other aspects of migration for a given species represent adaptive
responses to existing conditions. Natural selection is likely to act quickly on
nonadaptive behaviors. We do not need to look at the migration pattern of a
species as though it were a piece of sedimentary rock formed through eons of
slow change. Migration is a dynamic process, surely subject to rapid evolutionary change. If this is so, then the dramatic differences that we see among the existing forms of migration are quite possibly the result of different selective
forces acting on different sets of gene pools (adaptations). We need not expect
that a single explanation of migration must cover all observations. This conclusion is helpful, because several of the ideas that have been proposed seem
to explain movements of particular species but are less applicable to others.

Ancient Environmental Changes

Advancing glaciers, moving continents, and fluctuating Tertiary sea levels,
among other geological events, have all been proposed as the impetus to migration (Wallace 1874; Gräser 1905; Wolfson 1948). According to these theories, the sedentary ancestors of migratory birds were originally pushed (south
usually) by a massive environmental change. They returned when they
could—the following spring or, in some cases, generations later—because of
"ancestral habit" (Wallace 1874; Gräser 1905; Wolfson 1948). There are at
least two problems with these theories. First, some of the geological changes
suggested as providing impetus for the original migration, such as continental
drift, took place long before most of the species, and probably most of the
modern genera of birds, existed. Second, the inheritance of an ancestral habit
implies the maintenance of a genetically based behavior (e.g., homing to original breeding sites that were ice-covered for hundreds or thousands of years)
for generations in the absence of positive selection forces.

Availability of Resources Elsewhere

In theories based on the availability of food or some other resource (such as
safe nesting sites or mates) at some distance from the point of origin, birds mi-

grate to capitalize on the availability of this resource (Allen 1880; Taverner 1904; Pycraft 1910). This idea is a classic example of Lamarckism, as Gauthreaux (1982) has correctly pointed out. No evidence indicates that individuals within a population can evolve behavioral adaptations to suit a particular environment, any more than giraffes evolved long necks because of the existence of tall trees. However, many of these kinds of theories can be related to modern theories of seasonal resource use by simply inserting a few words about potentially increased fitness benefits to dispersing individuals.

Proximate Factors

Migration has also been theorized to be the result of an immediate response to harbingers of environmental change, such as decreasing temperature, shortening days, or changes in barometric pressure (Marek 1906; Walter 1908). These theories confuse the apparent triggers of migration (proximate causes) with the actual selection forces favoring the evolution of migration (ultimate causes) (Lack 1954). Although these factors could be the same under certain circumstances, it seems probable that such factors as changes in photoperiod could serve as appropriate cues for initiation of migratory movement, but they are unlikely to be ultimate causes of such movement.

Climatic Changes

The idea that climatic change plays a major role in the origin of migration is very old (Aristotle, Pliny). Cohen (1967:15) perhaps has provided the most cogent argument for the evolution of migratory behavior in populations confronted with climatic change:

Regular seasonal migration may now be considered as a special case of the more general behavioral pattern of migration in response to environmental conditions which are positively correlated with subsequent low survival for a nonmigrating population. The regularity of seasonal migration can be explained as being the result of an adaptation to a combination of a regularly low expectation of survival for a nonmigrating population, and of a high correlation between the conditions at the time the decision to migrate is being made, and subsequent survival.

Cohen's argument that the individual's reduced probability of survival in temperate winters favors southward movement makes a great deal of sense for species that clearly are adapted to allow them a choice between migrating or remaining on a temperate breeding ground during the winter (e.g., parids and fringillids). It is more difficult to fit the model to obviously tropical species,

like the hummingbirds, or to intratropical and austral migrants in which failure to move from their tropical breeding sites does not appear to threaten their
survival. The model also does not take into account the possibility that movement could be related to maximizing reproduction rather than survival.

Seasonal Use of Fruit or Nectar Resources

Levey and Stiles (1992:447) have proposed that "seasonal movements within the tropics predisposed [resident frugivores or nectarivores] to migration
out of the tropics." They note the increasing evidence of widespread intratropical movements, particularly among frugivorous species. Members of
these species—for example, the Resplendent Quetzal—move considerable
distances over the course of an annual cycle, apparently tracking the seasonally variable distribution of preferred fruits in different habitats (Powell and
Bjork 1994). The hypothesis of Levey and Stiles has considerable appeal as an
explanation for the migratory movements of some species; however, the majority of Nearctic–Neotropical migrants are not wholly, or even principally,
frugivorous or nectarivorous. As discussed in Chapter 3, migrants are quite diverse in their food use habits. Indeed, fully a third of the species feed on
aquatic animal or plant material, whereas another large segment forages
mainly on terrestrial animals.

Seasonality and Interspecific Competition

Cox (1985) proposed a hypothesis to explain the origin of long-distance
Nearctic–Neotropical migration based on the prior existence of populations at
an intermediate latitude. This model was based on earlier ideas of his own
(Cox 1968) plus more recent ideas and information on migrants (e.g., Greenberg 1980). Cox (1985:471) stated:

An integrated time-allocation and competition theory is suggested, in which the Mexican Plateau and arid southwestern United States act as a staging area for the evolution
of migration. In this region, where environmental unpredictability favors the reduction
of site tenacity, various forces select for the movement of birds into neighboring areas
with more favorable seasonal conditions, creating partial migrants. The increased seasonality of climate, the speciation of geographically isolated migrant population segments, and the elimination of permanent resident population segments by interspecific
competition lead to disjunct breeding and nonbreeding ranges.

This hypothesis may fit for some southwestern steppe species but is less appealing for species wintering in the highlands, wetlands, or moist lowlands of

the Tropics. In particular, several species have long-distance migrant and tropical resident populations that winter in the same region (e.g., the *Vireo olivaceus* superspecies complex).

Seasonal Change and Dominance Interactions

Gauthreaux (1978, 1982) has proposed that migration evolves as a result of seasonal change in resource availability at the point of origin. This hypothesis is similar to that presented by Cohen (1967), but it includes additional predictions regarding the results of intraspecific competition. Gauthreaux's model predicts that different classes of individuals within the migrating population will move different distances based on their relative abilities in intraspecific competition for resources; that is, dominant individuals (adult males in most cases) migrate the shortest distances away from the breeding site, and subordinate individuals (females and immatures) migrate the farthest. This explanation coincides with some data on winter distribution of age and sex groups of some species of partial migrants in the temperate zone, but there are other logical explanations for these movements based on the different morphologies, physiological requirements, and roles of the different age and sex groups (Ketterson and Nolan 1983). Also, the theory does not seem applicable to species apparently derived from tropical resident populations (e.g, Cattle Egret).

Migration Threshold Hypothesis

Baker (1978), in an attempt to formulate a general theory to explain all types of animal migration, proposed that each organism has a genetic "migration threshold." When conditions deteriorate sufficiently at point of origin, the individual migrates. The term "conditions" in this sense applies not only to the availability of food resources but also to the probability of reproductive success. Costs of migration are factored in as well. The problem with Baker's hypothesis is that it is so broad that it provides little illumination of the process of evolution of migration in specific instances. Nor does it explain why an animal should return to point of origin.

Summary

Most of the above theories strive to present a single explanation for the origin of migration. In examining the various types of movement patterns, however, it seems clear that no one explanation is sufficient, primarily because the competitive environment is different for different species. For some species, year-round residence in the temperate zone is a possibility, providing the opportunity for trade-offs between reproduction and survival. For such species, the ideas of Cohen (1967) may provide a sound explanation; for birds of highly seasonal, arid zones of the subtropics, the proposals of Cox (1985) seem sound; and for movements of tropical frugivores, the hypothesis of Levey and Stiles (1992) is logical. But we do not believe that any of these ideas adequately explains the evident movement of large numbers of tropical species from relatively aseasonal environments into temperate regions to breed. In the following section, we present our explanation of how this phenomenon could have evolved, and we provide citations in support of each proposed step.

HYPOTHESIS FOR THE ORIGIN OF MIGRATION IN SEDENTARY TROPICAL SPECIES IN RELATIVELY ASEASONAL ENVIRONMENTS

We propose that, for many Nearctic–Neotropical migrants, migration is a behavior that has developed principally in tropical resident species in response to pressure on young individuals to locate suitable, uncontested feeding and breeding habitat. This hypothesis has been put forward many times before as the southern ancestral home theory (Dixon 1897; Mayr and Meise 1930; Williams 1958; see summary in Gauthreaux 1982), but without the factual links necessary to provide the logical, selection-based steps from sedentary tropical resident to long-distance Nearctic migrant. Most of the factual links are now available, although some are on firmer ground than others. We present the hypothesis in stepwise fashion below to demonstrate where the ideas are supported by data and where further observation and testing are needed to evaluate the particulars.

Most students of migration have assumed that, in order for migration to occur, something must change at the point of origin. This observation does not really tell us very much. Probably no places on earth are stable in the sense that the competitive environment is the same for all individuals throughout an annual cycle. Even in a completely stable physical environment, more offspring will be produced than the environment can accommodate. Thus the act of reproduction itself causes change in the environment. What happens to these excess individuals? Many studies of numerous different taxa have documented what happens to them: They move. Sometimes they move short distances, and sometimes long distances. But in even the most sedentary populations, the offspring move (Fogden 1972; Willis 1974).

Based on the assumption that some members of a population of even sedentary species will move, several factors could favor the development of long-distance movement to a Nearctic breeding area for members of certain

Neotropical species under particular conditions, as outlined in the following scenario:

1. Because of intense intraspecific competition and high predation rates in populations of many tropical avian species, annual productivity is low (Fogden 1972; Willis 1974; Skutch 1976; but see Oniki 1979).
2. Despite low productivity, surviving young exceed the amount of space available for breeding because of the longevity of adults (Fogden 1972; Willis 1974; but see Karr et al. 1990).
3. Under intense intraspecific pressure, young individuals newly driven from parental territories are forced to move long distances in search of available habitat in which they can compete successfully for mates and space. In resident birds, these movements favor evolution of physiological and behavioral characteristics normally associated with migration, such as hyperphagia, subcutaneous fat reserves, and homing ability (Ramos and Rappole 1994).
4. Some species exploit microhabitats (e.g., leaf litter or canopy) that are common to many habitat types, allowing them the option of at least trying other areas in different habitats or regions (Rappole and Warner 1980; Stiles 1980).
5. Increased availability of food resources, reduced competition, and reduced predation rates during summer allow much higher reproductive rates in temperate populations relative to populations of the same species in the Tropics (Ricklefs 1972; Skutch 1976). Individuals able to reach these temperate habitats have the potential for achieving significantly higher fitness rates than their tropical relatives (MacArthur 1972).
6. Once the new migrant of tropical origin completes its breeding at a new site, it should remain unless the environment deteriorates. If the environment deteriorates, it should return to its point of origin in the Tropics, for two reasons: The point of origin is a known area of suitable habitat, and it is already "programmed in" for homing purposes. Similarly, the new migrant should return to its Nearctic breeding site when competition for breeding sites recurs in the annual cycle.
7. If the new breeding environments allow much higher production rates than those at the point of origin, offspring produced in the new environments will eventually flood the winter habitats, outcompeting the resident breeders by putting more new individuals into the system than the resident breeders do.

Step 6 is obviously the most fragile link in the development of a hypothesis to explain long-distance migration. It is easy enough to argue the selective value to an individual of having the capability to leave a site in the face of pressure (from either the physical or the social environment). But once the bird has moved, why should it come back? Early theorists, unencumbered by having to provide a sound argument based on natural selection, saw the logic of southern ancestral home theories quite readily. Birds left the Tropics to exploit resources in the temperate zone, where they could produce more offspring at lower risk of predation. That these theorists had no mechanisms to account for why an individual should make this a round-trip was not considered a problem.

Most recent theorists have preferred northern ancestral home theories because it is easier to make a selection-based argument that the birds were originally pushed south in fall by resource scarcity and gradually returned north in spring as the weather allowed. Even these arguments require a certain suspension of rigor in order to account for the return of birds to the breeding ground, particularly young individuals. If a group of birds has moved southward to a wintering site quite suitable for survival, and that wintering site only gets better in terms of resources as the seasons progress from winter to spring to summer, what pushes the bird northward to return to an area where it will have to compete for breeding territories?

Most current explanations of migration ignore the question of why an individual should make a return trip once it has moved away, evidently presuming that this individual will automatically return at some later time to its breeding site. There is no a priori reason to expect the bird to return to its breeding site, however, if conditions at the wintering site are acceptable. Yet, for the phenomenon to be migration, as opposed to dispersal, not only must something force the individual bird away from its point of origin initially, but something must also push it back to its point of origin later in the annual cycle. Also, this something must act over and over again.

In our hypothesis, the factor that forces the first bird away from its tropical point of origin is competition with conspecifics for breeding space. What pushes it back is deterioration of weather in the temperate zone. Movement at both ends of the journey can be explained in terms of an individual's behavior in response to changes in specific aspects of its environment. Timing, routes, and triggering mechanisms become modified by natural selection over generations. Thus, we believe that the weakest point of our hypothesis is not the question of what causes our tropical paradigm to move northward in spring or southward in fall. Rather, the weakest point is the evident enormity of the social, physiological, and ecological obstacles to success for the first individual making such a journey. Based on our current understanding of migration, these obstacles seemingly imply a very low probability of success, unless the

Table 6.1

Species with coexisting breeding resident and wintering migrant populations in the Neotropics

Species	Locality and investigator(s)
Pelecanus occidentalis	Lesser Antilles (Leopold 1963)
	Netherlands Antilles (Voous 1955)
Ixobrychus exilis	Costa Rica (Dickerman 1971)
	Panama (Eisenmann and Loftin 1968)
Ardea herodias	Panama (Eisenmann and Loftin 1968)
	Lesser Antilles (Leopold 1963)
	Montserrat (Danforth 1939)
Butorides striatus	Panama (Eisenmann and Loftin 1968)
	Hispaniola (Wetmore and Swales 1931)
	Lesser Antilles (Leopold 1963)
Nycticorax nycticorax	Hispaniola (Wetmore and Swales 1931)
Cathartes aura	Panama (Eisenmann and Loftin 1968)
Pandion haliaetus	Venezuela (Friedmann and Smith 1950)
Elanoides forficatus	Panama (Eisenmann and Loftin 1968)
Buteo jamaicensis	Panama (Eisenmann and Loftin 1968)
Porphyrula martinica	Jamaica (Cruz 1972)
Gallinula chloropus	Panama (Eisenmann and Loftin 1968)
Fulica americana	Hispaniola (Wetmore and Swales 1931)
Charadrius wilsonia	Panama (Eisenmann and Loftin 1968)
	Netherlands Antilles (Voous 1955)
C. vociferus	Puerto Rico (Danforth 1937)
	Hispaniola (Wetmore and Swales 1931)
Himantopus mexicanus	Panama (Eisenmann and Loftin 1968)
Catoptrophorus semipalmatus	Lesser Antilles (Leopold 1963)
Gallinago gallinago	Venezuela (Friedmann and Smith 1950)
Sterna maxima	Lesser Antilles (Leopold 1963)
S. hirundo	Lesser Antilles (Leopold 1963)
	Netherlands Antilles (Voous 1955)
S. fuscata	Panama (Eisenmann and Loftin 1968)
Rynchops niger	Panama (Eisenmann and Loftin 1968)
Zenaida asiatica	Nicaragua, Guatemala, Honduras (Blankinship et al. 1972)
Z. macroura	Panama (Eisenmann and Loftin 1968)
Coccyzus americanus	Hispaniola (Wetmore and Swales 1931)
	St. John Island, U.S. Virgin Islands (Robertson 1962)
Chordeiles acutipennis	Panama (Eisenmann and Loftin 1968)
C. minor	Panama (Eisenmann and Loftin 1968)
	Hispaniola (Wetmore and Swales 1931)
	Lesser Antilles (Leopold 1963)

Species	Locality and investigator(s)
Pyrocephalus rubinus	Veracruz, Mexico (Brodkorb 1948)
Progne subis	Panama (Eisenmann and Loftin 1968)
	Netherlands Antilles (Voous 1955)
	Guyana (Snyder 1966)
Vireo olivaceus	Hispaniola (Wetmore and Swales 1931)
	Venezuela (Friedmann and Smith 1950)
	Guyana (Snyder 1966)
Dendroica petechia	Panama (Eisenmann and Loftin 1968)
	Venezuela (Friedmann and Smith 1950)

process evolved in tiny steps over long periods of geological time. An alternative to long periods of time, however, is large numbers of individuals giving this option a try.

Three lines of evidence lend support to our southern ancestral home hypothesis: the taxonomic relationships of Nearctic migrants, examples of rapid development of long-distance migration, and examples of continual movement of tropical species into temperate regions. These sources of evidence are discussed below.

Migrant Systematic Affinities

One of the principal sources of evidence supporting a tropical origin for many Nearctic–Neotropical migrants derives from their taxonomic relationships. These relationships indicate that these species are Neotropical birds that happen to breed in the temperate zone (Dixon 1897; Mayr and Meise 1930; Mayr 1946; Williams 1958; Cox 1968; Rappole and Warner 1980; Smith 1980; Stiles 1980; Rappole and Tipton 1992). Most of these species appear to derive from Neotropical forms, which would help explain why they are able to exploit resources in ancient and complex Neotropical environments (Mayr 1946; Mengel 1970; Barlow 1980). Seventy-eight percent of all Nearctic migrant species have congeners or conspecifics that currently have breeding populations in the Neotropics (see Appendix 1). Several species even have permanent populations of breeding residents and winter residents coexisting in the same tropical community or region during certain seasons of the year (Table 6.1). Thus the question of why migrants do not breed in the Neotropics, which has so intrigued biologists (Hamilton 1962; Klopfer et al. 1974), is answered by the fact that 48% of Nearctic migrant species do breed in the Tropics (see Appendix 1).

It could be argued that the existence of Neotropical breeding populations for 162 migrant species is not the result of the invasion of tropical species into temperate habitats but the reverse, that is, the invasion of temperate species into tropical environments. This argument runs contrary to much of ecological community theory, which holds that tropical communities are older and more complex than temperate communities, and therefore tropical niches are likely to be filled by highly specialized tropical residents (see Chapter 4). Nevertheless, this explanation is at least possible, and it is even attractive for certain groups of migrants to the Tropics that appear, because of their taxonomic affinities and/or habitat preferences, to be temperate in origin. For some of these groups, including such genera as *Toxostoma, Melospiza, Regulus, Pipilo,* and *Carduelis,* an origin in the subtropical montane or desert areas, following the scenario proposed by Cox (1985), seems possible. Invasion into the tropical region by members of these species could have occurred as the result of climatic changes to arid or highland regions in the northern Neotropics, which created new habitats similar to those in the southern temperate region.

A temperate origin is more difficult to defend for groups that have large numbers of tropical breeding representatives or that are widespread as breeders throughout major portions of the Neotropics. Such a situation presumably would reflect a long history of tropical occupation. Yet many migrant taxa show this type of distribution. The genus *Polioptila,* for instance, is represented by 1 species of long-distance migrant to the temperate zone, 8 species resident in Middle or South America, and 1 Cuban species (Fig. 6.1). Similarly, the genus *Vireo* is represented by 11 species that breed north of the Mexican border and 32 that are resident in the Tropics (Barlow 1980), and *Icterus* by 5 species of Nearctic migrants and 18 species of tropical residents. Thirty-five of the 39 families that include Nearctic migrants have as many or more tropical-breeding species as they have temperate-breeding species (Rappole and Tipton 1992).

Rapid Development of Nearctic–Neotropical Migration

Given the biological pressures in the Tropics, the preadaptations for migration that many tropical species possess, and the availability of abundant resources and reduced predation pressure in the temperate zone during the summer, evolution of a migratory habit seems inevitable (Rappole and Tipton 1992). That evolution of migration is an ongoing process is demonstrated not only by the existence of so many species with both migrant and tropical resident populations but also by data on species like the Cattle Egret, in which the development of Nearctic migratory populations has occurred within the past 40 years.

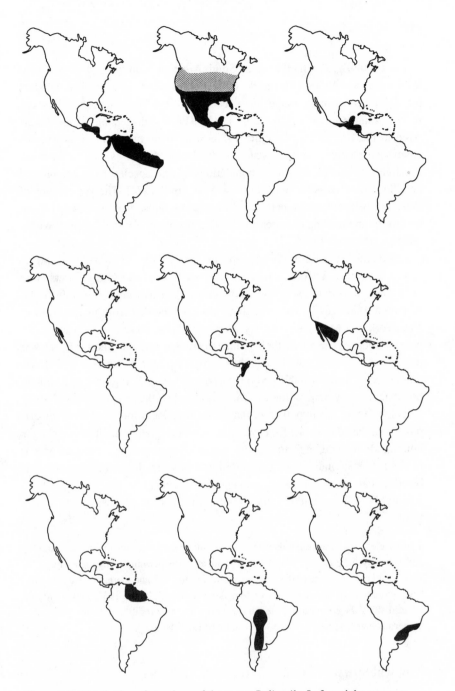

Figure 6.1. Distribution of members of the genus *Polioptila*. Left to right, top row; *P. plumbea, P. caerulea* (shaded area indicates breeding range), *P. albiloris.* Middle row: *P. nigriceps, P. schistaceigula, P. melanura.* Bottom row: *P. guianensis, P. dumicola, P. lactea.* The range for the Cuban representative of the group, *P. lembeyei,* is not depicted.

The Cattle Egret, which invaded South America from the Old World Tropics in the late 1800s (Meyer de Schauensee 1966), is a tropical resident throughout most of its range. First records for the Western Hemisphere are from Suriname in 1877 (Gladstone 1983). The first dispersing wanderers arrived in the United States around 1952. Since then, members of the species have been recorded in most states of the United States as well as in many Canadian provinces. Northern populations of the species have established breeding colonies as far north as New York (Bull 1974). The populations of these northern-breeding egrets are Nearctic migrants, arriving in early summer to set up breeding colonies and leaving in autumn for their tropical winter quarters (Peterson 1980).

Nor is the Cattle Egret the only species that has recently developed migratory populations. Tricolored Herons, Snowy Egrets, and Great Egrets were once Nearctic migrants in the northeastern United States (Bartram 1791; Audubon 1840–1844) with breeding colonies recorded from as far north as New York and Massachusetts. By the early 1900s, these species were extirpated throughout most of their temperate range, and the Snowy Egret was nearly extinct in the continental United States (Eaton 1910; Bull 1974). Yet these species have reestablished themselves as Nearctic migrants and summer breeders in many parts of their former range in less than a century (Bull 1974; Peterson 1980). We hypothesize that tropical resident forms of these species possessed the adaptations for development of a migratory habit (hyperphagia, homing ability, and so forth). Development of migration in these instances is likely to be an individual, behavioral response to the availability of new breeding habitat, rather than to some genetic change.

Bull (1974:74) provided a graphic description of this process for the Snowy Egret:

This species, more than any other heron, was valued for its ornamental plumes and consequently suffered severely. It had become almost exterminated by the early 1900's and only because of the strictest protection in 1913 and thereafter escaped possible extinction. At any rate it started to make a comeback by the 1930s and greatly increased along the coast during the next two decades. As a result, it is today the most numerous breeding heron on the south shore of Long Island.

Tropical Species Found in Temperate Habitats

The process of range expansion and development of migratory populations by tropical species is dynamic and continuous. This phenomenon should not be surprising, because the movement of these birds is probably nothing more than dispersal away from the breeding site. That there is always pressure forcing dispersal by individuals of even species that are poorly adapted for temperate environments is evidenced by the number of tropical species reported from the temperate zone each year (Table 6.2). The occurrence of these birds hundreds or even thousands of kilometers away from their normal range may be not the expression of a particular genetic trait for movement but merely the coincidence of random movement away from the parental territory and the occurrence of favorable resource concentrations at intermediate stops.

POSSIBLE WINTERING GROUND INFLUENCE ON MIGRANT EVOLUTION

The concept of tropical origins for many Nearctic migrants forces a major shift in zoogeographical thinking about these species. Most attention relating to speciation and centers of evolution for Nearctic migrants has focused on breeding ground distribution and on climatic events that might fragment or otherwise affect that distribution (Rand 1948; Larson 1957; Mengel 1964, 1970; Hubbard 1973; Greenberg 1979; Johnson and Zink 1985; Bermingham et al. 1992). However, as several authors have pointed out (e.g., Lack 1944; Salomonsen 1955; Fretwell 1972; Rappole and Warner 1980; Rappole and Tipton 1992), competition for food on the wintering ground may be the most likely determinant of individual survival for many migrant species. Therefore, it may play an important role in the evolution of Nearctic migrants, affecting not only their adaptation but their wintering distributions and perhaps species formation as well.

Maps and tables in Rappole et al. (1993b: Appendixes 4 and 5) show current distribution and status information for Nearctic migrants. These data are presented here by political divisions in Figure 6.2. Figure 6.3 summarizes data on winter distribution by major geographical regions, indicating that the greatest concentrations of transient and/or wintering migrants occur in the countries of Middle America, particularly Mexico. These summaries are important in terms of where conservation efforts need to be directed, but they tend to obscure zoogeographic relationships that exist between breeding and wintering areas in many groups of migrants. These relationships are well illustrated by analysis of some of the distributional patterns in the wood warbler subfamily Parulinae.

Mengel (1964) examined patterns of breeding distribution for several groups of warblers and found an apparent correlation between their present ranges and the probable glacial events of the Pleistocene. He did not consider

Table 6.2
Some North American distributional records for tropical species

Species	Locality in North America and investigator(s)	"Normal" range
Buteogallus anthracinus	Florida (Abramson 1976)	West Indies, Mexico, and Central America
Columbina inca	Oklahoma (Bartinicki 1979)	Southern Arizona, Texas, Mexico, and Middle America
Crotophaga ani	Georgia (Moore 1974)	West Indies and Florida
C. sulcirostris	Oklahoma (Delap 1979) and Colorado (Webb 1976)	The Neotropics to South Texas
Amazilia violiceps	California (Johnson and Ziegler 1978)	Mexico
Eugenes fulgens	Kansas (Boyd 1978)	Southern Arizona, Mexico, and Central America
Heliomaster constantii	United States (Witzeman 1979)	Mexico and Central America
Calypte costae	Canada (MacKenzie-Grieve and Tatum 1974)	Southeastern United States and Mexico
Euptilotis neoxenus	United States (Zimmerman 1978)	Mexico
Empidonomus varius	United States (Abbott and Finch 1978)	South America
Tyrannus melancholicus	Florida (Ayers et al. 1980)	South Texas, southern Arizona, and the Neotropics
T. dominicensis	Mississippi (Weber and Jackson 1977)	West Indies
T. savana	Maryland (Wierenga 1978), Wisconsin (Freese 1979), and South Carolina (Dick 1974)	The Neotropics
Ridgwayia pinicola	United States (Wolf 1978)	Mexico
Quiscalus mexicanus	Illinois (Bohlen 1976) and Colorado (Stepney 1975)	Southeastern United States to northern South America

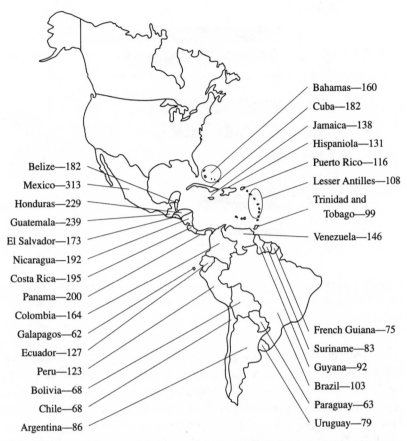

Bahamas—160
Cuba—182
Jamaica—138
Hispaniola—131
Puerto Rico—116
Lesser Antilles—108
Trinidad and
   Tobago—99
Venezuela—146

Belize—182
Mexico—313
Honduras—229
Guatemala—239
El Salvador—173
Nicaragua—192
Costa Rica—195
Panama—200
Colombia—164
Galapagos—62
Ecuador—127
Peru—123
Bolivia—68
Chile—68
Argentina—86

French Guiana—75
Suriname—83
Guyana—92
Brazil—103
Paraguay—63
Uruguay—79

Figure 6.2. Total number of migrant species recorded in the various political divisions of the Neotropics. The number for Hispaniola includes species recorded either as transients or as winter residents.

winter ranges of these species as relevant to the speciation process. However, maps showing both the breeding and wintering ranges for two of the groups that Mengel considered reveal some remarkable parallels among breeding location, winter range location, and pattern of species distribution (Figs. 6.4 and 6.5). Each group contains a single species with a broad breeding range across the northeastern Nearctic and an extensive winter range in the Caribbean lowlands of Middle America. In contrast, each group also contains several species that have breeding ranges arranged on a more or less north–south axis in the mountains of western North America and that winter in relatively restricted pockets in the mountains of northwestern Middle America. Bermingham et al.

Region

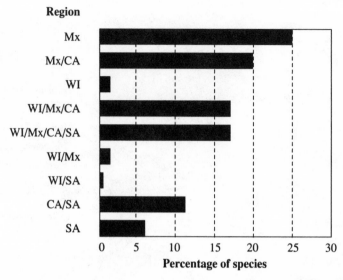

Figure 6.3. Percentage of migrant species wintering in major geographical regions. Mx = Mexico, WI = West Indies, CA = Central America, SA = South America.

(1992) have questioned the precise pattern of relationships delineated by Mengel (1964), but they appear to accept the basic premise that all that is required for speciation to occur is isolation of breeding populations for a prolonged period. If that were so, however, why would there be such extensive speciation in genera occupying the western United States (e.g., *Dendroica, Vermivora,* and *Empidonax*) but not in the eastern United States? Mountains occur in the east as well as in the west, with numerous opportunities in the Appalachian and Ozark regions for extensive isolation of breeding populations of spruce–fir–inhabiting species during interglacial periods, such as the present (Morain 1984).

Species assemblages of a number of avian groups show distributional patterns that are broadly similar to those seen in the wood warbler genera discussed by Mengel (1964): (1) Trochilidae, (2) *Empidonax* flycatchers, (3) *Tyrannus* flycatchers, (4) *Myiarchus* flycatchers, and (5) the subgenus *Vireo* (see maps in Rappole et al. 1993b: Appendix 4). Each of these groups shows considerable differentiation among western taxa and relatively little differentiation in the east. Such similarities in species distribution for a number of unrelated groups are difficult to attribute to chance. We suggest three possible explanations: (1) The degree of breeding ground isolation has entirely controlled the speciation process, as in the hypothesis proposed by Men-

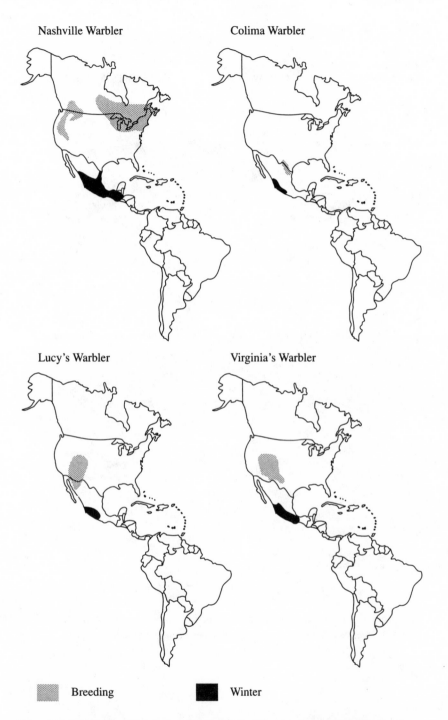

Nashville Warbler

Colima Warbler

Lucy's Warbler

Virginia's Warbler

Breeding   Winter

Figure 6.4. Breeding and wintering ground distribution of members of the Nashville Warbler (*Vermivora ruficapilla*) superspecies complex.

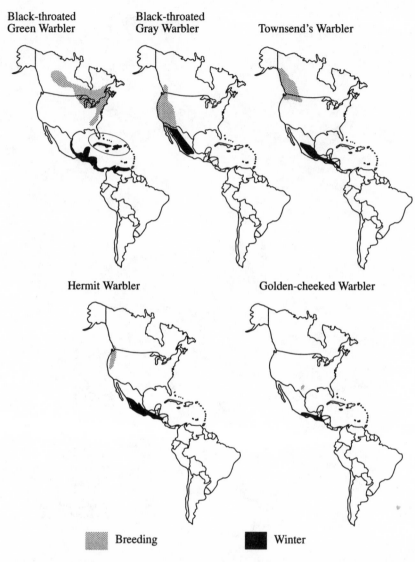

Black-throated
Green Warbler

Black-throated
Gray Warbler

Townsend's Warbler

Hermit Warbler

Golden-cheeked Warbler

Breeding   Winter

Figure 6.5. Breeding and wintering ground distribution of members of the Black-throated Green Warbler (*Dendroica virens*) superspecies complex.

gel (1964) and Bermingham et al. (1992); (2) changes in both breeding and wintering grounds affected evolution of these species' groups; and (3) the species originated as sedentary populations on the wintering grounds, which were isolated by Pleistocene events, and evolved a migratory habit at a later time.

It is not possible to differentiate among these possibilities at present, and of course there are other possibilities as well, which may become clearer as our understanding of plant distributions during glacial and interglacial periods grows and our knowledge of the time of population isolation for the avian taxa investigated also grows. However, the parallels between the potential isolation of both breeding and wintering sites (as occurs often in the west during interglacial periods) and the splintering of some western groups may indicate that a relationship exists between the degrees of simultaneous isolation of breeding and wintering areas, and that it has some bearing on the speciation process. Perhaps, for speciation to occur, populations of a migrant species must be isolated on both breeding and wintering areas—on the breeding area because genetic isolation is necessary, and on the wintering area because intense competition for resources between members of two different populations would ultimately lead to extinction or competitive exclusion of individuals from one of the populations. Any genetic change in a population is likely to affect the competitive ability of its members relative to that of the members of another population. Theoretically, two populations of a species geographically isolated on the breeding ground could be forced by competition to remain genetically similar if they overlapped in a large portion of their winter ranges, because a change in one of the populations might affect the existing competitive balance between members of the different populations. This explanation could account for lack of speciation in birds like the Nashville Warbler. Its eastern and western breeding populations are presumed to have been isolated for a minimum of 11,000 years (Mengel 1964), the period for which a gap in the eastern and western distribution of temperate coniferous habitat is presumed to have existed. Perhaps the isolated breeding populations of eastern and western Nashville Warblers have failed to differentiate because their winter ranges overlap considerably (see Figure 6.4).

Ample opportunity for simultaneous isolation of breeding and wintering ground populations of migratory birds occurred in the western United States, western Mexico, and the West Indies during the Pleistocene, when habitat islands or actual oceanic islands were common (Morain 1984; Van Devender 1990). As populations of a species, isolated from each other on their breeding ground, became progressively different, members of one population may have excluded members of another from various parts of their winter range. Thus we hypothesize that competition and wintering ground distribution may have played and may continue to play important roles in the evolution of migrant species.

SUMMARY

Eight categories of explanations have been provided for the evolution of migration: ancient environmental changes; availability of resources elsewhere; proximate factors such as photoperiod or temperature changes; climatic changes; seasonal tracking of fruit or nectar; seasonality and interspecific competition; seasonal change and intraspecific dominance interactions; and Baker's migration threshold hypothesis. None of these explanations seems to provide a cogent, natural selection–based argument for evolution of northward migration into the Nearctic by tropical nonfrugivores, a common category of long-distance migrants.

We propose a hypothesis to explain evolution of migration in such species, in which migration results from intense intraspecific competition for breeding sites. Though several parts of the hypothesis are supported by available information, some portions are based on speculation. However, nearly all parts of the hypothesis can be tested by examining and comparing the life history characteristics of species in which temperate and tropical breeding populations continue to coexist, as in the Red-eyed Vireo (*Vireo olivaceus*) and many other migrants. The principal evidence supporting this hypothesis is the extensive taxonomic information documenting the existence of close tropical resident relatives for most long-distance migrants (78%) to the Nearctic.

The possibility of a tropical origin for a significant number of these temperate-breeding taxa forces reevaluation of the significance of the tropical portion of the life cycle for these birds. Most analyses of the evolutionary history of these taxa have ignored the winter distribution and the distribution of tropical relatives. We propose that these distributions need to be considered if a complete understanding of the zoogeography and evolution of Nearctic–Neotropical migrants is to be understood. Many migrants apparently originated from tropical species. Their annual return to the Tropics has profound effects on their evolution and speciation. Any consideration of migrants, zoogeographic or otherwise, must consider competitive interactions not only on the breeding ground but during migration and on the wintering ground as well.

7

OLD WORLD VERSUS NEW WORLD MIGRATION SYSTEMS

Each regional pattern of bird migration is unique, reflecting its own particular historical, biogeographic, and geological peculiarities. Nevertheless, there is a tendency to use the exegesis developed for one region to explain the system of another, especially when data are lacking to provide a complete picture. Thus, ideas formed from analysis of the Palearctic–African migration system, in particular the work of Morel and Bourlière (1962) and Moreau (1972), have helped shape understanding of the Nearctic–Neotropical system. In this chapter, we review the Palearctic–African and Palearctic–Asian systems; we consider the concepts of migrant ecology and evolution that have been developed based on the Old World system, particularly the European–African system; and we contrast those concepts with concepts derived from our studies of the New World system and the studies of McClure (1974) and others on Asian migrations.

Studies by Cooke (1915) of New World migrations provided the beginning of a biogeographic concept of a migration system, but his work lacked the information on tropical environments, movements, and distributions necessary for a coherent hemispheric viewpoint. The remarkable work of Moreau (1972) provided that viewpoint for the Palearctic–African system. Moreau listed 185 species of birds as migrants from the Palearctic into Africa, joining the 1,480 native African species (see Appendix 2).

CHARACTERISTICS OF NEOTROPICAL, AFRICAN, AND ASIAN MIGRATION

Perhaps the most striking aspect of the list of Palearctic–African migrants (Appendix 2), in comparing it with the lists of Nearctic migrants (Appendix 1) and the Asian migrants (Appendix 3), is that there are far fewer Palearctic–African migrants than Nearctic or Asian migrants. Moreau's list did not include birds restricted to coastal environments, a group that constitutes about

20 species on our list of Nearctic migrants. But even excepting that group, the number of Nearctic–Neotropical migrants is 72% larger than the number of Palearctic–African migrants (318 and 185 species, respectively). The Asian migration system, however, is very similar to the Nearctic– Neotropical system, at least in terms of numbers of species. Though the Asian system is perhaps not as well known as the New World or Palearctic–African systems, because of a lack of extensive international banding efforts, 338 Asian species are known to migrate from the Palearctic to the Tropics of southern India, Southeast Asia, the Sunda Islands, and the Philippines (see Appendix 3). Some possible explanations for the numbers of migrant species in the different regions are discussed below.

Availability of Forest Habitat

A major difference among the world's migration systems is the relative amount of forest available to a potential migrant in the Tropics, the subtropics, and the temperate areas. As Table 7.1 illustrates, the Palearctic–African system is characterized by a virtual absence of forest-related migrant species (Moreau 1952; Karr 1976a; Monkkonen et al. 1992). Roughly one-third (112 species) of all Nearctic migrants, and very nearly the same number of Asian migrants (107 species), are forest-related during the breeding season (see Chapter 2 and Appendix 3), whereas Moreau (1972) reports a total of 48 species of forest-related Palearctic–African migrants (44 species, according to Peterson et al. 1967—see Appendix 2). However, of those Palearctic–African migrant species that use forests during the breeding season, the majority inhabit open country during the winter in Africa. Monkkonen et al. (1992) reported that of all the Palearctic–African migrant passerines that use forests during the breeding period, only 3 species are found principally in broad-leaved forest habitats on their wintering grounds in Africa: Pied Flycatcher (*Ficedula hypoleuca*), Collared Flycatcher (*F. albicollis*), and Wood Warbler (*Phylloscopus sibilatrix*). Our analysis of Palearctic migrant habitat use in Africa (see Appendix 2) provides a slightly larger number of species that winter in forest. Part of the difference has to do with different definitions of what is forest. Nevertheless, the differences between the number of migrants that use forest on the wintering ground in Palearctic–African system and the number in either the Nearctic–Neotropical or the Asian migration system are striking. In the latter systems, most of the migrant species that breed in woodlands also winter in woodlands (see Appendix 3; Rappole et al. 1993b: Appendix 2).

Table 7.1
Numbers of wintering migrant species from the Holarctic region and resident species in tropical evergreen forests

Locality	Forest type	Residents[a]	Migrants	Investigator(s)
Africa				
Mont Nimba, Guinea	Equatorial rain forest	150	0	Brosset (1984)
Nigeria	Equatorial rain forest	?	8	Elgood et al. (1966)
Gabon	Equatorial rain forest	?	0	Brosset (1968)
Lamto, Ivory Coast	Equatorial rain forest	113	0	Thiollay (1970a,b)
Various localities	Equatorial rain forest	?	6	Moreau (1972)
Central America				
Panama	Young-growth (>50 yr) and old-growth (>100 yr) wet lowland forest	131	17	Willis (1980)
Costa Rica	Lowland rain forest	149	20	Stiles (1976); Powell et al. (1992)
Mexico	Lowland rain forest	120	17	Andrle (1964); Rappole and Warner (1980)

Source: Rappole (1991).
[a]? indicates that the actual number of resident species in these forests was not stated. However, reported numbers of resident species found at comparable sites are 150 to 200 species (Brown et al. 1982; Colston and Curry-Lindahl 1986).

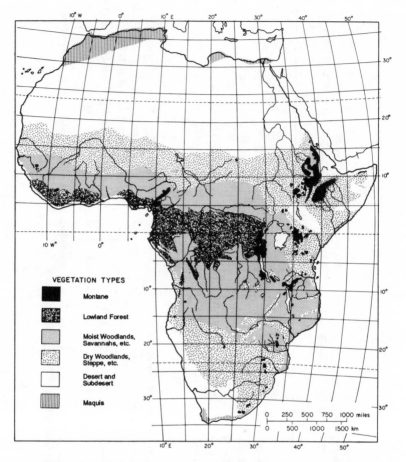

Figure 7.1. African vegetation types. Adapted from Moreau (1972).

In part, the lack of forest-related Palearctic–African migrants may be due to difficulties involved with tropical species attempting to invade European temperate forests, a feat that requires crossing 3,000 km of desert, maquis, and the Mediterranean Sea. As Figure 7.1 illustrates, forested habitats are virtually absent as potential stopover areas from central Africa to southern Europe (Rappole 1991). If migration evolves in species that can exploit microhabitats in a variety of habitat types connecting potential breeding and wintering sites, then absence of intervening forest habitat between those sites would be a significant barrier to evolution of a long-distance migration by forest-related species, regardless of the potential fitness benefits available to a successful migrant.

Diversity of the Subtropical Environment

Table 7.2 shows that the differences among Palearctic–African, Nearctic–Neotropical, and Palearctic–Asian migration systems are not due to the relative absence of forest-related species alone. There are 54 families of birds that contain one or more migratory species in at least one of the three migration systems. Thirty-six of these families are shared between the Nearctic–Neotropical and Palearctic–African regions (including both migrant and resident species) (Austin 1985). For 22 of these 36 families, there are more migrant species in the Nearctic–Neotropical region; 10 families have more migrant species in the Palearctic–African region; and 4 families have equal numbers of migrant species in both regions. Clearly, lack of woodlands is not the only factor affecting development of migration in the Palearctic–African region. A second important difference between these two migration systems is that the African subtropics represent a relatively severe environment when compared with either the Asian or the New World subtropics.

We define the northern subtropics as those regions between the tropic of Cancer and latitude 30° N. About 355 species of birds breed in the New World subtropics. At least 85 of those species, and probably considerably more (undocumented as yet because of a lack of banding work), are partial or complete migrants to the Tropics during the nonbreeding period (Edwards 1972; Rappole and Blacklock 1985; Rappole et al. n.d.). The northern subtropics of the New World are quite diverse, including extensive forest, scrub, and aquatic habitats. In contrast, most of the African subtropics is covered by desert, and accordingly, the numbers of species in the bird communities of this region are relatively low. Only one species of Palearctic migrant, the Sooty Falcon (*Falco concolor*), has a significant portion of its breeding range within the African subtropics. For most Palearctic–African migrants, the majority of their breeding distribution is to the north or east. Indeed, fewer than 50 species have any portion of their range within the region. The Asian subtropics, like the New World subtropics, is quite rich in terms of habitats and has a similarly diverse resident and migratory avifauna (Ali and Ripley 1968–1974; Lekagul 1968; Wildash 1968; McClure 1974; King and Dickinson 1975).

We have proposed that many species of Nearctic migrants are tropical species that took advantage, in an evolutionary sense, of reduced intraspecific competition, reduced predation, and increased resources to breed in the temperate zone (see Chapter 6). If this hypothesis is correct, a subtropical region rich in hospitable habitats could serve as a major driving force behind the evolution of bird movements between tropical and temperate regions, providing corridors of favorable habitat between possible breeding sites. Lack of such a

Table 7.2
Numbers of species by family for Nearctic–Neotropical, Palearctic–African, and
Palearctic–Asian migration systems

Family	Nearctic–Neotropical species	Palearctic–African species	Palearctic–Asian species
Podicipedidae	4	0	1
Pelecanidae	2	0	2
Phalacrocoracidae	2	0	1
Anhingidae	1	0	0
Ardeidae	12	8	16
Threskiornithidae	4	3	4
Ciconiidae	1	2	6
Anatidae	20	11	19
Cathartidae	2	0[a]	0[a]
Accipitridae	11	17	23
Falconidae	4	8	6
Phasianidae	0	1	1
Rallidae	7	5	8
Gruidae	2	2	2
Jacanidae	0	0	1
Rostratulidae	0[b]	0	1
Burhinidae	0	1	1
Glareolidae	0[b]	2	1
Charadriidae	8	7	11
Haematopodidae	1	0	1
Recurvirostridae	2	2	2
Scolopacidae	30	20	38
Laridae	21	5	17
Columbidae	5	1	2
Cuculidae	3	3	9
Strigidae	3	2	2
Caprimulgidae	5	3	1
Apodidae	4	4	4
Trochilidae	13	0[a]	0[a]
Trogonidae	1	0[a]	0[a]
Upupidae	0[b]	1	1
Alcedinidae	1	0	2
Meropidae	0[b]	1	2
Coraciidae	0[b]	1	1
Picidae	3	1	1
Tyrannidae	32	0[a]	0[a]

Family	Nearctic– Neotropical species	Palearctic– African species	Palearctic– Asian species
Alaudidae	0	2	2
Hirundinidae	8	5	5
Troglodytidae	3	0	0
Campephagidae	0[b]	0	5
Pycnonotidae	0[b]	0	1
Dicruridae	0[b]	0	4
Oriolidae	0[b]	1	4
Muscicapidae	12	50	97
Mimidae	2	0[a]	0[a]
Motacillidae	2	6	12
Bombycillidae	1	0	0
Laniidae	1	6	3
Sturnidae	0	0	6
Zosteropidae	0[b]	0	2
Vireonidae	11	0[a]	0[a]
Emberizidae	92	3	6
Fringillidae	2	0	4
Ploceidae	0	1	0
Total	338	185	338

Source: Based on Appendixes 1–3.
[a]Family does not occur in Old World.
[b]Family does not occur in New World.

region would presumably serve as a serious barrier, at least to some kinds of species, such as birds of forested habitats.

MIGRANT TAXONOMIC RELATIONSHIPS

Of the 185 species of Palearctic–African migrants, 42 species (23%) have conspecific populations that breed in the African Tropics, and another 97 species (52%) have congeners that breed in that region (see Appendix 2), for a total of 139 species of Palearctic–African migrants that have close relatives that breed in the Tropics. Of the 338 species in the Asian migration system, 106 (31%) have conspecific populations that breed in the Tropics, and another 156 (46%) have congeners that breed in the region (see Appendix 3), for a total of 262 species of Asian migrants with close relatives that breed in the

Tropics. Though the proportion of migrants with conspecifics that breed in the Tropics is much higher in the Nearctic system (48%) than the proportions in the other migrant systems, the percentages of migrant species with close tropical relatives (congeners plus conspecifics) are very similar for all three systems (78% for both the Nearctic and Asian migrants, 75% for the Palearctic–African migrants). These numbers indicate that, as is the case for the New World system, a significant percentage of European and Asian birds are derived from tropical species. The persistence of more breeding populations of conspecifics in the New World Tropics may indicate a more recent occurrence of speciation in that region.

At least as interesting as the differences among the three major Northern Hemisphere migration systems are the similarities. In all three cases, a large number of apparently tropical species have invaded the temperate zone, and in all three cases one group has undergone an apparently recent explosion in species diversity, filling the small avian insectivore niche in temperate forests. In the New World, that group is the emberizid subfamily of wood warblers (Parulinae), with 51 species of migrants. In the Old World, that group is the muscicapid subfamily of Old World warblers (Sylviinae), with 29 species in the Palearctic–African system and 45 species in the Asian system.

THE MIGRANT ROLE IN DIFFERENT TROPICAL MIGRATION SYSTEMS

Current thought regarding the role of Palearctic migrants, as opposed to that of tropical African residents inhabiting the same communities, has recently been summarized well by Leisler (1990). He presents the following generalizations in contrasting migrants with tropical resident species, recognizing that there are likely to be species that do not necessarily fit this pattern:

1. Migrants prefer drier, more open, and peripheral habitats.
2. Migrants prefer a wider range of habitats.
3. Migrants are subordinate to residents.
4. Migrants are opportunistic in terms of resource use.
5. Migrants are less able to use mobile or hidden prey.
6. Migrant differ morphologically. They are smaller (but see Moreau 1972), are more generalized in structure, and have shorter bills and more attenuated wings.

These generalizations derive from extensive work by a number of individuals (see review in Leisler 1990), including Morel and Bourlière (1962),

Moreau (1972), Karr (1976), and Lack (1986). Most of their conclusions, in turn, were based on study of migrants in the tropical savannah communities of central Africa. These communities are idiosyncratic, in that they are highly seasonal; even many of the "resident" species occupying them are nomadic to a greater or lesser extent (Morel and Bourlière 1962). The question that arises from this work is whether the comparisons reflect fundamental aspects of the competitive relationship between "northern migrants" and "tropical residents" or whether they represent differences between one subset of tropical (migrant) savannah inhabitants and another, nonmigratory subset. Are these generalizations applicable to migrants in Asia and the Western Hemisphere, or are they restricted in their application to a specific group of Palearctic–African migrants? We will now consider this question on a point-by-point basis.

1. Migrants prefer drier, more open, and peripheral habitats. Of the species in our list of Nearctic migrants, 112 species winter in tropical forests of various kinds, and 110 species of Asian migrants use forested habitats on their Asian wintering grounds. Similarly, many austral migrants in South America are forest-related (Chesser 1994).

2. Migrants prefer a wider range of habitats. The number of habitats that are used by a species may have more to do with *when* population regulation is occurring during the annual cycle than with some innate difference between "migrants" and "residents" (see Chapter 8). If less habitat is available on the wintering ground than an incoming population of migrants requires, then use of a wide range of less suitable habitats would be an obvious result.

3. Migrants are subordinate to residents. The evidence indicating migrant subordinancy to residents is derived principally from observations at resource concentrations. Away from such concentrations, evidence of any kind of interaction has been slight to nonexistent (Chipley 1976; Powell 1980; Rappole et al. 1983, 1993b; but see Leisler 1990). Even in cases in which residents have been documented to supplant migrants (e.g., Leck 1972a; Leisler 1990), no evidence of the effects of these interactions on migrant survivorship has been presented.

Leisler (1990:167) has described a situation involving evident direct competition between migrants and residents in central Africa: "In our turdid guild some wintering species, which are subordinate to residents, can cope with the dominant local species by making more flights and by frequently changing perches. This restless/fugitive strategy allows the migrants to squeeze in, or even to superimpose their territories over that of others; to pounce to the ground; and to escape the attacks of the dominant species."

Migrant subordinancy is part of an argument designed to explain how migrants can fit into complex tropical communities in which resident species are likely to competitively exclude them from use of critical resources. If, however, migrant fitness is unaffected by resident activity when both groups are using the same set of resources, then the types of interactions, strategies, or tactics involved may be unimportant.

4. Migrants are opportunistic generalists in terms of resource use. Certainly many individuals, especially young birds or transients (Stutchbury 1994), of both migrant and resident species are nomadic and opportunistic, taking a wide variety of foods. However, many individuals in both groups are not resource generalists (Greenberg 1986, 1987a). The majority of individuals in a significant number of migrant species have been found to be territorial, to show site faithfulness throughout a season, and to return to the same wintering site in subsequent seasons in both the Nearctic–Neotropical system and, where studied, the Asian system (Nisbet and Medway 1972; Rappole et al. 1983, 1993b).

5. Migrants are less able to use mobile or hidden prey. This concept was developed by Thiollay (1988), based on his comparison of migrant and resident foliage gleaners in the Neotropics. In addition to the Old World hypothesis that migrants simply skim the most easily harvested resource, other explanations are possible. The resources used by residents may differ from those used by migrants because "residents" are, by definition, those species whose distribution is restricted to the Tropics. That is, they are residents because the resources they are adapted to exploit do not exist in the subtropics or in temperate regions. Thus, the average resource use patterns of residents as a group may differ from migrants as a group, not because residents deny specialized niches to migrants but because migrants use a different set of niches from that of residents, niches that are available in the subtropics and temperate regions as well as in the Tropics.

6. Migrants differ morphologically. The morphological differences that have been documented between migrants and residents fall into two categories: adaptations likely to be related to the making of regular long-distance flights (e.g., smaller size and more attenuated wings) and adaptations related to the migrant's ecological role in the tropical avian community (e.g., more generalized structure and shorter bill). Examples of studies in which morphological comparisons of migrant and resident groups have been performed include those by Cox (1968), Herrera (1978), Greenberg (1981a), and Leisler (1990).

Greenberg (1981a:147), in his study comparing foliage-gleaning tropical residents with foliage-gleaning migrants concluded: "The bill morphology of these [migrant] species supports the hypothesis that it is breeding resources that most strongly influence the shape of foliage-gleaning bills." He further speculated that the reason that the migrant species examined had not established breeding populations in the Tropics was that they were unable to compete with tropical residents for larger Orthoptera, which are presumed to be the critical resource during the tropical breeding period for foliage gleaners.

Greenberg's basic conclusions seem sound—that members of the chosen group of tropical residents (foliage gleaners) have, on average, longer, narrower bills than members of the migrant group, and that this bill structure is superior to a shorter, broader bill for handling larger insect prey such as Orthoptera. However, Greenberg's broader speculations regarding migrant inability to compete for critical resources in tropical communities are questionable for the following reasons. First, the groups of migrants and tropical residents compared are from taxonomically diverse assemblages: The migrants are from 5 genera representing 2 families (Vireonidae and Emberizidae); the tropical residents are from 13 genera representing 4 families (Formicariidae, Vireonidae, Vireolaniidae, and Emberizidae). No members of the two groups are from the same genus, and half of the resident species are from a family that has no migrant representatives. As discussed in Chapter 4, this type of study raises the issue of whether the comparison groups differ solely in the presumed question at issue (preferred prey items) or whether phylogenetic considerations could confound the analysis. Second, it has not been demonstrated that Orthoptera are critical for all foliage gleaners during breeding. Third, three of the migrant species included in the analysis are members of superspecies complexes in which one or more members have breeding populations in the Tropics (*Vireo olivaceus, V. gilvus,* and *Parula americana*), and all of the migrants included have congeners that breed in the Tropics. Thus, whatever the significance of bill shape differences between such groups as antbirds and warblers, it is not necessarily related to their relative abilities to use critical (breeding period) resources in tropical communities.

As with Greenberg (1981a), a difficulty with a number of the studies contrasting morphological differences between migrants and residents is that although the structural comparisons are quite specific and straightforward, the conclusions are often broad, reaching beyond the basic assumptions of the analysis. For example, when a comparison is made between migrants and residents in which resident curassows (*Crax rubra*), pheasant cuckoos (*Dromococcyx phasianellus*), oropendolas (*Gymnostinops*), and antshrikes (*Thamnophilus* spp.) are compared with migrant species (Cox 1968), we must ask what the discovered differences mean in terms of fundamental ecological relationships. Do the discovered differences tell us that migrants are fugitive species, or simply that there are certain groups of residents—for instance, large frugivores or ant followers—whose resource requirements cannot be met on a long-distance journey through the subtropics to the temperate zone?

One of the interesting aspects of the Palearctic–African migration system is the evident invasion of African savannah species into European forests. This occurrence has been explained as the result of the plasticity of migrants; that is, in order to be a migrant, a bird must be a generalist. Another explanation is possible, however. If these birds are viewed as tropical savannah species to begin with, we can hypothesize that they flew north originally to reduce intraspecific competition, occupying habitats for which they were not truly adapted. They were able to succeed because they had few competitors. The tropical, forest-related, foliage-gleaning species could not make the journey because there weren't enough forests along the route. The savannah species were successful invaders because there was enough intervening savannah and open habitat to allow them to make the journey.

SUMMARY

The Palearctic–African system of avian migration is the most studied of the world's migration systems. The species involved, the timing and routes of movements, and the breeding and wintering distributions are known with much greater accuracy than is true for migrants from any other region. In addition, ornithologists have been studying the ecology, behavior, and interactions of migrants in this system for considerably longer than in other migration systems. As a result, the ideas of pioneers such as Moreau, Morel, and Bourlière have had enormous influence on how migration is understood in other parts of the world. However, the Palearctic–African migration system has unique features that may invalidate its use as a model for understanding the development and function of other systems. These features include a lack of forested environments and a lack of environmental diversity in the subtropical region of Africa relative to Asian, South American (austral), and Nearctic–Neotropical systems.

Several characteristics have been attributed to migrants principally as a result of studies of migrant-resident interactions observed in the savannahs of central Africa. These characteristics have been formalized as a "migrant strategy" and have been used to understand the role of migrants wherever they occur. This strategy may be summarized as a procedure whereby a group of temperate species "fit" themselves into an ecologically packed tropical avian community. But these attributes may apply only in the idiosyncratic situations in which they were observed. Neither the characteristics nor the overall strategy appear appropriate for understanding the development and function of other migration systems.

To understand other migration systems, it may be more helpful to view migration as a filtering process, with the tropical bird community providing the material to be filtered. One type of filter would be the material with which the process begins. Evolution in the New World Tropics provides different material for the migration filter than in the African Tropics or the Asian Tropics. The filtering process is reflected in the families from which the migrants of each region are derived—similarities and differences depend on the material that was available at each site. A second filter is the variety of subtropical habitats available. These habitats provide the stepping stones for tropical species to spread northward into the temperate zone. A third filter is the variety of temperate zone habitats available. A fourth filter may be the occurrence of breeding and wintering sites isolated from conspecifics, which could allow rapid speciation (Lack 1944; Salomonsen 1955; Rappole et al. 1983, 1993b).

That significant morphological and ecological differences exist between major groupings of migrant and resident species should not be a surprise, given the kinds of resources and habitats available to a potential migrant. Care needs to be taken, however, in interpreting these differences as anything other than a reflection of the success enjoyed by one particular group in exploiting a new habitat as opposed to the inability of another to do so.

8

MIGRANT POPULATION CHANGE

MIGRANT VULNERABILITY

Some species of migratory birds may be more vulnerable to extinction than either tropical or temperate resident species. For some biologists, however, this prediction is counterintuitive. Migrants are generally considered less vulnerable to extinction than residents because migrants have multiple habitat options: breeding, postbreeding, stopover, and wintering. But in the case of some migratory species, these options may instead represent fragile links in a chain of successively necessary habitats that are required to complete an annual cycle. Such species may depend on three or more geographically separate sets of habitats in order to survive an annual cycle rather than depending on only one set of habitats.

Consider the following model based on resident and migratory species with similar ecological requirements. Assume that the annual likelihood of extinction due to catastrophic habitat alteration is equal to .001 in each of three sets of habitats. The annual probability of extinction for a resident species in any one of these habitats is .001. The probability of extinction for a migratory species that depends on all three habitats during different portions of the annual cycle is equal to the combined probabilities of extinction for all three habitats, which would be .002997, nearly three times the probability of extinction for a given resident species.

In approaching this problem, we assumed that although a migrant is exposed to each habitat for only a portion of the annual cycle, the catastrophic alteration of the habitat could occur at any time after the migrant's departure the previous year up to and including the migrant's stay during the next year. Thus the total annual probability of .001 would apply to the migrant in each habitat as well as to a resident. The combined probability is then calculated in the same way as calculating the probability of throwing at least one head in three separate coin tosses. That is, the probability (P) of throwing one head in three tries is added to the probability of throwing two heads in three tries, plus the probability of throwing three heads in three tries:

$$P \text{ (event will occur once)} = \frac{3!}{1! \, (3-1)!} \, x^1 \, (1-x)^{3-1}$$

$$P \text{ (event will occur twice)} = \frac{3!}{2! \, (3-2)!} \, x^2 \, (1-x)^{3-2}$$

$$P \text{ (event will occur 3 times)} = \frac{3!}{3! \, (3-3)!} \, x^3 \, (1-x)^{3-3}$$

where x is the probability that the event will occur once. In our migrant habitat scenario, "catastrophic alteration of habitat" includes any change that makes the habitat no longer suitable—for example, a new predator, disease, or competitor for the same resources, or the complete destruction of the habitat, as in the conversion of forest to pasture.

HISTORICAL ASPECTS

A generation ago, the potential for serious decreases in Nearctic migrant populations was noted by Carson (1962), Aldrich and Robbins (1970), Vogt (1970), and a few others (see review in Rappole et al. 1983). These authors suggested that declines could be caused by pesticides, destruction of breeding habitat, destruction of wintering habitat, changes in agricultural practices, and a number of other factors. Evidence that such declines not only were possible but also were, in fact, occurring was based mostly on anecdotes or data gathered from local surveys.

Data from the first 15 years of the U.S. and Canadian breeding bird surveys were equivocal in their indications. Some species showed regional declines (e.g., Prairie Warbler, Loggerhead Shrike, and Sora), but most Nearctic migrants showed no significant trends or even showed increases in some cases, despite some local long-term surveys that revealed significant decreases (Robbins et al. 1986). C. S. Robbins (1980:33), who has been the guiding force behind the Breeding Bird Survey since its inception in 1966, discussed these species' differences, contradictory trends, and fluctuations:

Actually, many different factors are acting on our long-distance migrants and it is not always possible to determine the relative importance of these factors. For example, we believe that the recent increase in many of our warblers in the eastern United States is a result of recovery from extensive use of persistent pesticides during the 1950's and early 1960's. Data from a much longer period are needed to determine relative effects of loss of breeding habitat, loss of winter habitat, and change in pesticide and other pollution levels.

One problem with continent-wide surveys, like the Breeding Bird Survey, is that they suffer from some procedural problems that reduce their usefulness in reflecting population changes in Nearctic migrants. Variation in observer efficiency, natural cycles of population fluctuation correlated with multiyear environmental cycles, sampling methods, methods of data treatment, or misunderstanding of breeding ground population dynamics all potentially affect the reliability of the surveys with regard to relative numbers of birds in a population over time. Despite these problems, the Breeding Bird Survey has two significant advantages: It has been done systematically throughout the country, and it has been done in the same way every year for the past 28 years. Therefore, it is worthwhile to examine the Breeding Bird Survey information carefully to determine the strengths and weaknesses of the data set.

NATIONAL BREEDING BIRD SURVEY

Is there any way to determine whether a population decline has occurred? And if a decline is observed, what caused it? If a general, species-wide decline existed, it would probably be on the order of some part of a percentage point per year. Spread over all of the populations of a species, each of which could be fluctuating naturally from local environmental effects, such a gradual species-wide decline might be difficult to detect until the cumulative effects over a number of years were so large that they would be difficult to miss, even by the casual observer.

Part of the difficulty of detecting declines arises from the methodology of the only range-wide assessment of numbers that we have—the Breeding Bird Survey. In this survey, estimates of population changes are based on numbers of singing birds counted along established routes, with hundreds of volunteers across the country surveying the same routes each year. Song count accuracy is based on three assumptions: (1) that the observer is able to detect and identify all territorial birds based on their songs; (2) that each singing bird represents a male-female pair of adult, breeding individuals; and (3) that all males that are paired sing at the same rate.

Research has shown that these assumptions are not always valid, for the following reasons:

1. Timing the counts to be made when only breeding members of the community are present is quite difficult. In northern Virginia, for instance, transient individuals of the same species as the resident breeding bird community continue to pass through on migration until the end of May—and these birds sing. By July, many birds have completed nesting and are silent. The amount of song in these communities declines significantly through June to mid-July (Rappole et al. 1992; McShea et al. n.d.). Thus there is a narrow window of time, essentially the month of June for northern Virginia, in which song counts can be performed with any hope of obtaining an accurate count of breeding community members.

2. Males of supposedly monogamous species may have more than one mate. The percentage of occurrence of serial polygyny can vary significantly among species, and even among different populations of the same species, serving as a potentially serious bias in estimating sizes of breeding populations (Lack 1968b; Thompson and Nolan 1973; Rappole et al. 1977; Stewart et al. 1977; Powell and Jones 1978; Wittenberger 1980).

3. Unpaired birds (floaters) occur in significant numbers in some breeding populations (Darwin 1871; Hensley and Cope 1951; von Haartman 1971; Rappole et al. 1977, 1992).

4. Unpaired males may sing as much as or more than paired males in some species (Frankel and Baskett 1961; Nice 1964; Stone 1966; Krebs 1971; Armbruster et al. 1978; Baskett et al. 1978; Nolan 1978; Rappole and Waggerman 1986; Swanson 1989; Morton 1992).

5. Males of isolated pairs may sing less than mated males that live in denser populations (LaPerriere and Haugen 1972; Marion 1974; Sorola 1984).

6. Breeding Bird Survey routes are generally located along roads, where habitat change may occur more rapidly than in the general landscape.

7. "Analysis of population trends from 1966 through 1991 indicates that failure to include observers as covariables in the analysis [of Breeding Bird Survey data] results in an overly optimistic view of population trends" (Sauer et al. 1994:50).

The most serious of these biases is probably the singing by unpaired males. Initial effects of a rangewide population decline might be that more first-year males would sing on territories in the absence of mates and that fewer males would be polygynous. A decline might therefore have little detectable effect on the number of singing males, at least initially. Thus, broad surveys based on singing males might be slow to assess major population trends.

It can be argued that even if biases in the Breeding Bird Survey data exist, they are unimportant because these are trend data, the purpose of which is to distinguish long-term tendencies. The data are not intended to be used for accurate assessments of populations. The problem is that if song does not have the relationship to avian populations that is the basic assumption of the method (i.e., one singing bird is equivalent to one pair of breeding birds), then questions are raised regarding the meaning of the information. If song varies by age, habitat, pairing status, time of the season, weather, density, and so forth, then what does a change in numbers of songs recorded nationwide for a species mean? Fewer males? Increased fragmentation? Microhabitat changes along the routes? We don't know.

LOCAL AND REGIONAL SURVEYS

Several local and regional surveys of Nearctic migrant populations have also been performed. A characteristic of these surveys is that a local focus is likely to lead to a search for a local explanation in interpreting the results. An example is the study by Temple and Temple (1976) of 35 species for which abundance data covering a 40-year period were available. About half of the birds examined were Nearctic migrants. Population decreases were found in some of them (e.g., American Bittern, Marsh Wren, and Peregrine Falcon), and these decreases were attributed to specific local conditions—marsh drainage and pesticides in the cases cited. Only one Nearctic migrant, the Yellow-throated Vireo, showed a definite decrease that could not be easily explained. The authors suggested that the disappearance of the American elm (*Ulmus americana*) was responsible in this case.

Ascribing changes in populations to changes in local conditions is the norm for the following reasons. First, local conditions often do cause easily detectable changes in local breeding avifauna. Walcott (1974) and Aldrich and Coffin (1980) provided examples of declines in Nearctic migrants resulting from the local impact of urbanization. Lum (1978) recorded declines in shorebird populations related to "renovation" of a wetland site. Pesticides, forest fragmentation, strip mining, dumping of nuclear wastes, and pollutants are highly visible and do cause obvious problems for local birds. Second, the dependent relationship that a migrant bears to its stopover and wintering areas has not been clearly understood. These areas often are not recognized as the delicate links in the species' life cycle that they are. Third, stopover or wintering ground habitat destruction is not local in its effect on a breeding bird population. Wiping out the Wood Thrushes in southern Veracruz does not eliminate the Wood Thrushes in central Minnesota. The local effects are diluted over a large portion the species' breeding range (Ramos and Warner 1980).

CURRENT TRENDS

Significant declines have been recorded for 109 species of Nearctic migrants within the past decade. These declines have been documented by a number of methods: long-term breeding site studies, rangewide breeding bird surveys, studies of transient volume during migration, and regional breeding pair surveys. The studies on which the declines reported in Table 8.1 are based were made over variable time periods ranging from 10 to 50 years. Statistical procedures vary among the studies, but where analyses were performed, as in the case of the Breeding Bird Survey data, only those species that registered statistically significant declines are included in Table 8.1. For some species, declines have been recorded by several different methods. Though some researchers question the meaning of these reported declines, suggesting that they may be due in part to various biases and design flaws (Hutto 1988a; James et al. 1992), the evidence appears to be sufficient to be cause for concern. But if the declines are real, what are the causes?

EVALUATION OF CAUSES OF APPARENT MIGRANT DECLINES

Recent reviews of the apparent population declines in migratory species have been performed by Askins et al. (1990) and DeGraaf and Rappole (1995). They summarize information on four types of studies that have documented declines: (1) rangewide breeding bird surveys, (2) long-term studies at one or more breeding sites, (3) studies of transient volume during migration, and (4) regional breeding ground investigations. Based on the findings of these studies, the authors address eight possible causes of migrant population declines, which are discussed below.

Breeding Habitat Loss

Disappearance of breeding habitat is the most commonly cited agent for migratory bird declines (U.S. Fish and Wildlife Service 1987). In many instances, this focus on the 60- to 90-day period that a migrant spends on its

Table 8.1
Species of Neotropical migrants with reported declines

Species	Type of study (and source)[a]
Accipiter striatus	L (39)
Buteo platypterus	T (5,41)
B. swainsoni	R (36)
Falco sparverius	T (5)
Bartramia longicauda	R (8,27)
Numenius phaeopus	T (18)
Calidris alba	T (18)
Limnodromus griseus	T (18)
Larus pipixcan	B (32), S (32)
Coccyzus erythropthalmus	R (44), S (32)
C. americanus	B (32), L (7,29), S (32)
Asio flammeus	R (38)
Caprimulgus vociferus	S (32)
Archilochus colubris	L (7,29)
Contopus sordidulus	L (35)
C. virens	B (32), L (7,29,33), S (32)
Empidonax flaviventris	T (19)
E. virescens	L (7,29)
E. minimus	L (1,17), R (44), T (19)
Myiarchus crinitus	L (7,10,29)
Tyrannus forficatus	B (40)
Hirundo rustica	S (32)
Cistothorus platensis	L (39)
Polioptila caerulea	L (1,7)
Catharus fuscescens	B (32), L (2,7,23,24,33), S (32), T (19)
C. minimus	T (19)
C. ustulatus	L (3,14,17,25), R (17), T (19)
C. mustelinus	B (32), L (2,7,20,23,29,33), R (17,44), S (18), T (19)
Dumetella carolinensis	L (7), R (44), T (19)
Bombycilla cedrorum	R (44)
Vireo griseus	B (32), S (32)
V. solitarius	R (44)
V. flavifrons	L (1,7,20,29,33,39)
V. gilvus	L (35), R (44)
V. olivaceus	L (2,7,20,22,23,24,29,33)
Vermivora peregrina	S (32), T (21,37)
V. celata	T (37)
V. ruficapilla	T (19,37)
Parula americana	L (7,20,29), S (32), T (13)
Dendroica petechia	L (7), R (44)
D. pensylvanica	L (24), R (44), S (32), T (37)
D. magnolia	L (3,14,24), T (37)
D. tigrina	S (32), T (13,37)
D. caerulescens	L (3)
D. coronata	R (44), T (13,19,37)

Species	Type of study (and source)[a]
D. virens	L (2,3,24,33), S (32), T (37)
D. fusca	L (15,17,43), T (37)
D. pinus	R (44)
D. discolor	B (32)
D. palmarum	T (37)
D. castanea	S (32), T (37)
D. striata	S (32), T (13,37)
Mniotilta varia	L (2,7,23,24,29,33)
Setophaga ruticilla	L (1,7,20,23,24,29,33), R (44), S (32), T (19)
Helmitheros vermivorus	L (29,33), S (32)
Seiurus aurocapillus	L (1,2,3,7,20,23,24,29,33), S (32), T (19)
S. noveboracensis	L (24), T (19,37)
S. motacilla	L (7,29)
Oporornis formosus	B (32), L (7,20,29), S (32)
Geothlypis trichas	L (7,35), R (44), S (32), T (13)
Wilsonia citrina	L (2,7,20,23,24,29,33,43)
W. pusilla	L (35), T (19,37)
W. canadensis	L (3,24,43), S (32), T (19,37)
Icteria virens	B (32), L (35)
Piranga olivacea	L (1,3,7,24,29), R (44), S (32)
Pheucticus ludovicianus	L (3), R (44), S (32), T (19)
P. melanocephalus	L (35)
Passerina amoena	L (35), S (32)
P. cyanea	B (32), L (3,7), R (44), S (32)
P. ciris	B (32), S (32)
Spiza americana	B (32)
Pipilo erythrophthalmus	B (12), L (35), R (44), T (13,19)
Spizella passerina	T (19)
Dolichonyx oryzivorus	B (30,32), R (6)
Sturnella magna	B (12), R (44)
S. neglecta	B (12)
Icterus spurius	B (32)
I. galbula ("Baltimore Oriole")	R (1,44)
I. galbula ("Bullock's Oriole")	B (32)

Source: DeGraaf and Rappole (1995).

[a] B = long-term (>20 yr), rangewide Breeding Bird Survey counts; L = long-term (>20 yr) comparative study at one or more sites; R = regional studies; S = short-term (<20 yr), rangewide Breeding Bird Survey counts; T = counts of birds in transit between breeding and wintering grounds. Numbers in parentheses refer to the following citations, which document the reported declines: (1) Ambuel and Temple 1982; (2) Askins and Philbrick 1987; (3) Baird 1990; (4) Bartgis 1992; (5) Bednarz et al. 1990; (6) Bollinger and Gavin 1992; (7) Briggs and Criswell 1979; (8) Carter 1992; (9) Confer 1992; (10) Criswell 1975; (11) Droege 1991; (12) Hagan 1993; (13) Hagan et al. 1992; (14) Hall 1984a; (15) Hall 1984b; (16) Hamel 1992; (17) Holmes and Sherry 1988; (18) Howe et al. 1989; (19) Hussell et al. 1992; (20) Johnston and Winings 1987; (21) Jones 1986; (22) Kendeigh 1982; (23) Leck et al. 1981; (24) Litwin and Smith 1992; (25) Marshall 1988; (26) Novak 1992; (27) Osborne and Peterson 1984; (28) Peterson and Fichtel 1992; (29) Robbins 1979; (30) Robbins et al. 1986; (31) Robbins et al. 1992a; (32) Sauer and Droege 1992; (33) Serrao 1985; (34) Serrentino 1992; (35) Sharp 1985; (36) Steidl et al. 1991; (37) Stewart 1987; (38) Tate 1992; (39) Temple and Temple 1976; (40) Titus 1990; (41) Titus et al. 1990; (42) Via and Duffy 1992; (43) Wilcove 1983; (44) Witham and Hunter 1992.

breeding site is well justified by observed changes to the nesting habitat. Some waterfowl species, for instance, are conspicuously dependent on prairie pot-holes for successful breeding. This habitat has undergone a marked reduction as a result of farming practices, which has in turn caused severe reductions in productivity of these and other waterbirds (Novak 1992).

For another group of species, currently understood to be declining, the original breeding habitat was artificially increased during the period of settlement in the eastern United States. These species (e.g., Alder Flycatcher, Bachman's Sparrow, and Bewick's Wren), which breed in grassland, savannah, or scrub habitats, underwent extensive expansions in range and distribution as forests were converted to fields in the 1700s and 1800s. Reversion of open areas to woodlands has caused these birds to disappear from many parts of their ex-panded ranges and to be placed on several state endangered species lists, even though survival in their original, presettlement range may not have been threatened (LeGrand and Schneider 1992).

In a third group of species, declines have been attributed to changes in breeding habitat for which the evidence of such changes is equivocal at best. Members of this large group, which includes such taxa as Least Bell's Vireo and Cerulean Warbler (Goldwasser et al. 1980; Robbins et al. 1992a), require additional study of all phases of the life cycle to determine the point at which the population is being controlled. Although many habitats used by a species over the course of a year may be threatened, only one of them is likely to be a bottleneck that controls population size (Wiens 1977), and conservation ef-forts should focus on that one habitat.

Habitat Fragmentation

The effect on population dynamics for species whose habitats are broken from large blocks into small pieces of varying amounts of isolation has excited both theoretical and practical interest for some time (Arrenhius 1921; Gleason 1922; Cain and Castro 1959; MacArthur and Wilson 1967; Villard et al. 1993). Most of this attention has focused on breeding habitat (see review in Askins et al. 1990), although a few researchers have considered whether populations could be affected by fragmentation of transient or wintering habitat as well (Rappole and Morton 1985; Robbins et al. 1987, 1992b; D. Petit and L. Petit, pers. com.).

MacArthur and Wilson (1967) formalized study of populations in frag-mented habitats and dubbed their resulting synthesis the theory of island bio-geography. According to the theory, the number of species in a community is dependent on a balance between the relative rates of colonization and extinction. Those rates, in turn, are affected by the size of the area the community occupies and by its degree of isolation. The model predicts that in large sites with a low degree of isolation, an equilibrium between colonization and extinction will be established that is at the high end in terms of species richness, whereas small, isolated sites will show an equilibrium that is at the low end.

Numerous studies of breeding bird communities in large versus fragmented habitats of eastern North America have shown that fragmented habitats are indeed less diverse in general than similar, nonfragmented habitats (Askins et al. 1990). However, the reasons for these differences have not been clear. For instance, several studies have found that although size of the patch appeared to be positively correlated with species richness, degree of isolation did not have a significant impact on this parameter (Galli et al. 1976; Ambuel and Temple 1983; Askins et al. 1990; Freemark and Collins 1992). Robbins et al. (1989a) measured this "area effect" for 75 forest-nesting species by examining 469 sites in Maryland and Virginia. The researchers were able to calculate a precise site size at which fragmentation began to have an effect on population size for an individual species occupying the site. Villard et al. (1993:759) concluded from their study of Ovenbirds in Quebec and Ontario that "habitat fragmentation reduces pairing success by altering dispersal dynamics or habitat selection by females." Interestingly, studies of European breeding bird communities have not revealed significant "area effects" (Haila 1986).

In addition to the theoretical predictions of MacArthur and Wilson (1967), several other explanations for observed changes in breeding bird communities as an apparent result of fragmentation have been put forward, including the following: increased exposure in fragments to brood parasitism by Brownheaded Cowbirds; nest predation (Whitcomb 1977; Lynch and Whitcomb 1978; Brittingham and Temple 1983; Wilcove 1985; Temple and Cary 1988; Robinson 1992); loss of critical microhabitats more likely to be represented in larger habitat blocks (Robbins et al. 1989a); and increased interspecific competition between "generalist" edge species and "specialized" forest interior species (Whitcomb 1977; Butcher et al. 1981; Askins and Philbrick 1987).

Successional Changes to Breeding Ground Habitat

The amount of habitat change that has taken place on the North American continent since invasion by Europeans has been immense. Perhaps recent declines in migratory species are a by-product of the ebb and flow of the extensive successional changes caused by human land use in the region (Hall 1984a; Litwin 1986; Askins and Philbrick 1987; Litwin and Smith 1992).

Breeding Habitat Alteration by White-tailed Deer

From near extinction in many parts of eastern North America at the turn of the century, white-tailed deer (*Odocoileus virginianus*) have undergone an astonishing recovery (Warren 1991). McCabe and McCabe (1984) estimate populations of 3 to 4 deer per square kilometer during presettlement times. Current wild populations can exceed 10 times that number in areas protected from hunting, such as national parks (McShea and Rappole 1992). The impact of high deer populations on understory vegetation is obvious, causing overall reductions in vegetation density and changes in forest regeneration patterns (Alverson et al. 1988). The question arises as to whether these habitat alterations affect migratory birds as well (Baird 1990; McShea and Rappole 1992). Leimgruber et al. (1994) found a significant negative correlation between vegetation density and predation rates on artificial nests. Deer could also depress overall community productivity and cause long-term changes in plant community structure (McShea and Rappole 1992). At present, however, possible effects of deer or other browsers and grazers, such as cattle, on migratory birds are mainly speculative and have yet to be thoroughly tested.

Contaminant Poisoning

Rachel Carson (1962) raised the alarm regarding the pervasive damage to ecosystems in general and birds in particular caused by contaminants. The decline of a number of species—including several migrants, such as Brown Pelican, Peregrine Falcon (Risebrough and Peakall 1988), and Osprey—has been attributed to these substances. Bans in the United States of the most pernicious chemicals have resulted in stabilization or recovery for a number of species. However, potent pesticides continue to be used in most countries where Nearctic migrants occur as transients or winter residents, causing continued high contaminant levels in breeding populations of some species, particularly raptors and waterbirds (Lincer and Sherburne 1974; Henny et al. 1982).

Normal Population Fluctuation

For many of the 109 species whose populations have shown declines, the evidence of decline is quite recent, dating back only a decade or so (Sauer and Droege 1992). Populations for a number of these species showed no significant change or even showed increases in the decades prior to the 1980s (Robbins et al. 1986). Perhaps these changes are the result of natural climatic and/or prey cycles expected in a variable environment (Holmes et al. 1986; Blake et al. 1992; Erskine et al. 1992).

Procedural Biases

Measurement of population change in species whose breeding, stopover, and winter ranges extend over tens of thousands of square kilometers is a complex task. The difficulties of gathering, analyzing, and interpreting these data are well recognized and have led some researchers to question the validity of conclusions based on such data. James et al. (1992), for instance, have argued that apparent rangewide declines found by Robbins et al. (1989b) based on Breeding Bird Survey data were the result of incorrect statistical procedures. James et al. maintain that a different set of procedures reveals no significant, overall trend in population variation for these birds, but rather a combination of trends that vary according to region.

Hutto (1988a) has also challenged the conclusion that migratory birds as a group are showing significant declines. He points out that studies showing declines are much more likely to be published than studies that do not.

Stopover and Wintering Ground Habitat Loss

Destruction of certain habitat types in parts of the Neotropics has reached catastrophic proportions. Regional studies in Costa Rica, Mexico, and other portions of Middle America have shown lowland rain forest losses of 88% to 95% (Myers 1980a; Sader and Joyce 1988; Dirzo and Garcia 1992; World Resources Institute 1992; Rappole et al. 1994). Are these numbers important for migratory birds known to use these habitats as transients or winter residents? To some extent, the answer to this question lies in our understanding of the ecology and evolution of Nearctic migrants. If migrants are wandering interlopers from the temperate zone, exploiting temporary resource flushes in a variety of marginal habitats, then perhaps habitat destruction helps these species rather than harming them. However, if Nearctic migrants are derived from and members of Neotropical communities, with specific niches in these communities, then destruction of tropical habitats could threaten their populations as well as populations of tropical resident species. In addition to the problem of understanding the migrant's place in tropical communities, there is the added difficulty of determining when population regulation is occurring during the annual cycle. That is, even if the numbers do show population declines, there is no way at present to attribute those declines to events occurring at a specific time or place.

Rappole and McDonald (1994) have suggested a way to cut this Gordian knot—by making a set of predictions based on the likelihood that a population is controlled by breeding ground factors as opposed to wintering ground or stopover site factors. They propose that populations whose controlling factors

Table 8.2

Predictions and observations based on the hypothesis that populations of Nearctic migrants are controlled by breeding ground events

Predictions: If Nearctic migrant populations are controlled by breeding ground events, then . . .	Pertinent observations
1. Migrants should have their choice of optimal winter habitat sites and should not occupy suboptimal sites.	1. Migrants are often recorded in apparently suboptimal winter habitats (Karr 1976; Hutto 1980; Waide 1980; Lynch 1992).
2. Migrant breeding pairs should fill all available breeding space, including marginal breeding habitats.	2. Migrants often do not fill all available breeding space (Robbins et al. 1989a; Askins et al. 1990).
3. Migratory bird declines should not be observed in breeding habitats that are undisturbed and presumably optimal.	3. Declines have been observed in long-term studies of migrant bird communities on apparently undisturbed sites (Hall 1984a; Holmes and Sherry 1988; Marshall 1988).
4. Declines should not appear in species where no apparent change has occurred in breeding habitat.	4. Declines have occurred among forest-related migrants, despite apparent increases in amounts of total forest on the breeding ground (Birch and Wharton 1982; Powell and Rappole 1986).
5. Return rates of adults to breeding sites should be higher than returns to winter sites.	5. In the few studies done, the opposite has been found to be true (Rappole and Warner 1980; Holmes and Sherry 1992).
6. The number of young birds allowed to enter the breeding population should decrease as the amount of quality breeding habitat decreases relative to quality winter habitat.	6. In the few studies done, the number of young entering breeding populations has been increasing (Sherry and Holmes 1992).
7. The number of nonterritorial males (floaters) in the breeding population should be high.	7. In the few studies done, the number of nonbreeding males (floaters) in breeding populations has been declining (McDonald et al. n.d.).
8. The number of nonterritorial birds (floaters) in the winter population should be low.	8. Several studies at various locations have recorded high floater numbers in populations of wintering migrants (Rappole and Warner 1980; Rappole et al. 1989; Winker et al. 1990d; Wunderle 1992).

Predictions: If Nearctic migrant populations are controlled by breeding ground events, then . . .	Pertinent observations
9. Numbers of migrants in optimal breeding habitats should not fluctuate appreciably with natural cycles, because they will be buffered by the excess numbers of potential breeders in the population.	9. Breeding numbers have been found to vary sharply with natural cycles (Blake et al. 1992).
10. Numbers of territorial individuals in optimal winter habitat should show sharp annual fluctuations.	10. Where studied, wintering migratory bird densities in optimal habitats have remained remarkably stable across years and latitude (Winker et al.1990a), indicating the presence of large buffer populations in suboptimal habitats.
11. Declines in Nearctic migrants should be paralleled by changes in temperate, nonmigrant populations occupying the same breeding habitats.	11. Studies that have shown declines in migrant numbers have also shown steady or increasing numbers of temperate residents occupying the same habitat types (Lynch and Whitcomb 1978; Briggs and Criswell 1979; Askins et al. 1990; Blake et al. 1992; Litwin and Smith 1992).
12. Declines in Nearctic migrants should not be paralleled by changes in tropical residents occupying the same wintering habitats.	12. There are as yet no data on this question.

Source: Based on Rappole and McDonald (1994).

occur at different stages of the annual cycle are likely to have different characteristics and that these characteristics can be measured and even, in some cases, experimentally manipulated (Table 8.2). The authors conclude from their work that the characteristics for a large number Nearctic migrant populations indicate that wintering ground habitat loss is the single most probable cause of population declines.

SUMMARY

Migrants are especially vulnerable to extinction because of the multiple habitats they depend upon in different geographic locations at different times during the annual cycle. It has taken a long time for the environmental community to awaken to the threats confronting migrant populations, despite the early warnings provided in the 1960s by such prescient individuals as Rachel Carson and William Vogt. Part of the reason for the slow realization of the problems facing migrants has been the difficulty of determining migrant population trends. The national Breeding Bird Survey has been the principal method used for monitoring migratory bird populations. This method, as well as all other methods currently in use for this purpose, suffer from a number of procedural and interpretive difficulties. Nevertheless, nearly all types of surveys have now documented declines in some migrants, and nearly one-third of Neotropical migrants (109 species) have shown significant declines, according to one or more survey techniques.

Suggested causes for the observed declines include breeding habitat loss; habitat fragmentation; successional changes to breeding habitat; breeding habitat degradation caused by deer, cattle, or other herbivores; contaminant poisoning; normal population fluctuation; procedural biases; and stopover and wintering ground habitat loss in the Tropics. At present it is not possible to determine which of these problems, if any, is causing the observed declines, because even if a survey is accurate, it can tell us only whether the population is increasing or decreasing. It cannot reveal what is causing the change. We propose that it is possible at least to determine when during the annual cycle the population of a species is being limited. This period can be determined by making a series of predictions based on the characteristics of populations limited during different phases of the annual cycle. These characteristics can be tested for, providing potential clues to the causes of migrant population decline.

9

CONSERVATION

A principal purpose for summarizing information on Nearctic migrants in the Neotropics was to establish what conservation problems these birds confront when they leave their northern breeding areas. To what degree are migrants dependent on tropical stopover and wintering areas, and in what ways? Which regions and habitats in the Neotropics are most critical in terms of numbers of species potentially affected, and which areas are most threatened by environmental deterioration? Evidence has been presented in preceding chapters regarding the relationship between tropical environments and Nearctic migrants. Information on the ecology of many migrant species indicates that they may be limited by wintering ground competition. Therefore, what happens to their stopover and winter habitats is as important to their survival as what happens on their breeding areas. In this chapter, information is summarized on the present status of Neotropical environments and specific problems as they relate to migrants.

THREATS TO MIGRANT NEOTROPICAL HABITATS

Aquatic Habitats

Tropical ponds, marshes, rivers, lakes, estuaries, and coastal marine environments support 106 species of Nearctic migrants. These rich habitats are threatened by increasing pressures from changes associated with human population growth. These habitats are particularly vulnerable because wastes, pesticides, and other poisons resulting from abuse of surrounding habitats are carried by runoff into aquatic systems, where they concentrate and cause severe ecological damage.

Aquatic habitats and the organisms that live in them are part of a chain of cyclic breakdown, decay, and return of raw materials to the atmosphere that has continued virtually unchanged for millions of years. These cycles are now being disrupted in fundamental ways, mainly due to changes in the kinds and amounts of materials washed from the land: excessive amounts of soil, smothering and silting normal aquatic flora and fauna; toxic chemicals from agricultural lands and industrial dumps; and organic nutrients from urban centers (Ciflentes Lemus et al. 1971; De Goody 1971; Lara Madrid and Razetti 1971; Barducci 1972; Fernandez 1974; Pregnolatto et al. 1974; Homma et al. 1975; Lacombe and Moneiro 1975; Baez et al. 1976; Boyle Lemus et al. 1976; Branquinho and Robinson 1976; Dianese et al. 1976; ICAITI 1977; Julin and Sanders 1977; Roche 1977; Santos and Dos Pedrini 1978). These habitats are further degraded by oil spills, offshore dumping of garbage, and wetlands drainage and development for a variety of purposes, such as deepwater ports, tourism, and vacation homes (Rodriguez 1972; Canestri and Ruiz 1973; U.S. National Oceanic and Atmospheric Administration 1979; Roubal and Atlas 1980).

Little work has been done to date on the effects of aquatic habitat degradation on Nearctic migrants. The results of studies that have been done are not encouraging. As noted below in the section "Pollution," most of the birds that have been found suffering from pollutant poisoning are associated with aquatic ecosystems. As materials pass through the aquatic food chain, they become concentrated until they reach toxic levels in predators like the terns, herons, and falcons (Vermeer et al. 1974; Schlatter 1976; Albuquerque 1978; Figueroa et al. 1979; Henny et al. 1982, 1984; Henny and Herron 1989).

Shrub-Steppe Habitats

Many areas that fall into the shrub-steppe category are subject to extremes of temperature and precipitation that make them unsuitable for human development. As a result, desert, alpine regions, tundra, and the like remain more or less untouched, except where they are affected by global-scale environmental problems, such as acid rain, ozone depletion, or global warming. In addition to these sites, however, some of the potentially richest agricultural lands and pasturelands are in the shrub-steppe category, and most of them have been cleared and developed for food production decades ago, if not centuries ago. Pre-Columbian uses of these lands involved long-term cycles of shifting cultivation in which the land was allowed fallow periods to recover lost nutrients (Caufield 1984). The current combination of chemical and mechanical farming practices along with high human population densities has changed the intensity with which these lands are used, as well as the kinds and amounts of waste products generated by their use. As a result, pressures on prairies, savannahs, and shrublands have increased sharply in recent years, producing a series of now-familiar environmental problems: erosion, loss of soil fertility, soil compaction, salinization, and desertification (Brera and Shahrokhi 1978; Coiner 1980; Ollson 1985; Rappole and Ramos 1985; Ringrose and Matheson 1985). These problems have changed the biodiversity of such habitats. Furthermore, the application of pesticides, herbicides, and fertilizers causes the buildup of toxic materials in runoff and the eventual poisoning of even the groundwater (Lenon et al. 1972; Baez et al. 1976; Boyle Lemus et al. 1976; Branquinho and Robinson 1976; Companhía Técnica Saneamento Ambiental 1978).

Very little work has been done on the stopover or wintering ground natural history of shrub-steppe migrants. These migrants are among the species most often considered as benefiting from both tropical and temperate forest alteration, because their habitats presumably should be increased by forest reduction practices. Ironically, the U.S. Breeding Bird Survey and other counts show decreases for many of these birds—for example, Scissor-tailed Flycatcher, Bell's Vireo, Prairie Warbler, Yellow-breasted Chat, Loggerhead Shrike, Dickcissel, Painted Bunting, and Bobolink (U.S. Fish and Wildlife Service 1987; Titus 1990; Bollinger and Gavin 1992; Sauer and Droege 1992).

The evident decline of shrub-steppe migrants is most often attributed to factors associated with changes in agricultural practices in the United States (Bollinger and Gavin 1992). However, changes in the ways in which tropical agricultural habitats are used may be important as well.

The destruction of forest is visually dramatic, much more so than the destruction of prairies. Yet formerly forested regions are generally much less productive than prairies when converted to croplands. The shrub-steppe habitats are complex, diverse, and highly desirable in many instances for cultivation because rainfall and soil conditions are better suited for supporting a monoculture than forested regions are. As a result, there may be far less remaining undisturbed shrub-steppe habitat than forest habitat on a percentage basis, particularly in lowlands. Declines in shrub-steppe migrants may be related to this problem. The relationship between Nearctic migrants and shrub-steppe habitats needs to be investigated, and the amount of remaining natural shrub-steppe habitat of various kinds needs to be assessed.

Forest Habitats

A number of investigators and government agencies have examined the devastation of tropical forests on a world scale (e.g., Gomez-Pompa et al. 1972; Raven 1976; Sommer 1976; Nations and Nigh 1978; UNESCO 1978; Grainger 1980; Myers 1980a–c; U.S. State Department 1980; World Resources Institute

Table 9.1
Status of forests in the Neotropics

Country	Total area (km^2)	Amount presently forested (%)[a,b]	Annual rate of deforestation (% per year)[b]
Argentina	2,736,690	23	—
Belize	22,965	87	0.6
Bolivia	1,074,428	73	0.2
Brazil	8,456,510	80	0.5
Chile	748,800	23	0.7
Colombia	1,138,000	32	1.7
Costa Rica	49,132	45	3.6
Cuba	114,524	14	0.1
Dominican Republic	48,380	20	0.6
Ecuador	299,976	60	2.3
El Salvador	21,393	12	3.2
French Guiana	97,369	89	—
Guatemala	108,889	49	2.0
Guyana	215,000	87	0.0
Haiti	27,560	5	3.7
Honduras	112,044	63	2.3
Lesser Antilles	4,838	—	—
Mexico	1,193,133	33	1.3
Nicaragua	147,943	43	2.7
Panama	75,474	54	0.9
Paraguay	397,300	72	1.1
Peru	1,280,000	62	0.4
Puerto Rico	8,896	17	—
Suriname	162,000	89	0.0
Trinidad and Tobago	5,126	46	0.4
Uruguay	174,810	43	—
Venezuela	908,541	39	0.7

Note: — = no data available.
[a]Based on data from Myers (1980).
[b]Based on data from the World Resources Institute (1992).

1992). They have arrived at the startling conclusion that tropical forests are disappearing at the rate of 1% to 2% per year. Table 9.1 summarizes information on the current status of forest resources in countries of the Neotropics, based primarily on data published by the World Resources Institute (1992). Note that forests in the West Indies and Middle America, the most important migration stopover and wintering areas for migrants, are among the most significantly affected Neotropical forests in the hemisphere. The lowland moist forests are receiving the greatest pressure (Sader and Joyce 1988; Sader et al. 1991), but all major forest types are being reduced.

Forest loss on the order of magnitude reported above is difficult to imagine unless one is familiar with a specific area or region where the devastation has been observed in action. The Tuxtla Mountains of southern Veracruz are a microcosm of the process and can serve as a graphic illustration, both of the loss of forest and of the significance of that loss for migratory species.

The Tuxtla Mountains constitute a rugged, volcanic region roughly 3,000 km^2 in size, about 90 km southeast of the city of Veracruz. The region receives large amounts of orographic rainfall (2,500–4,500 mm annually, depending on elevation), which supported an abundant growth of rain forest (the "selva alta perenifolia" of Pennington and Sarukhan 1968) at the time of European settlement in 1532 (Medel y Alvarado 1963). The forests of the region were virtually untouched by indigenous human populations, who were forest-dwelling hunter-gatherers in the mountains (Foster 1942; Bernal 1969). As recently as the 1950s, an estimated 50% of the original rain forest remained in the region (Andrle 1964), but by 1986, satellite imagery revealed that less than 10% of the forest was left (Fig. 9.1) (Dirzo and Garcia 1992). Most of this land was cleared initially by subsistence farmers for crops but has subsequently been kept cleared as pasture for livestock (Fig. 9.2).

The Wood Thrush is a common Nearctic migrant in the Tuxtlas, where it occurs as a fall and spring transient and a winter resident. Banding and radio-tracking studies of the species in the Tuxtlas have revealed that Wood Thrushes hold individual territories in rain forest and various types of second growth throughout the winter period, remaining on territory throughout the winter season and returning to the same territory each year. The birds occur at densities of 2 to 3 birds per hectare in rain forest below 500 m in elevation and are scarce or absent at higher elevations or in open habitats (Rappole and Warner 1980; Rappole et al. 1989; Winker 1989). Of approximately 538 km^2 of forest that covered the northwest portion of the Tuxtlas below 500 m before European settlement, 28 km^2 (5.2%) is left. Consequently, the Wood Thrush is close to being extirpated in this region.

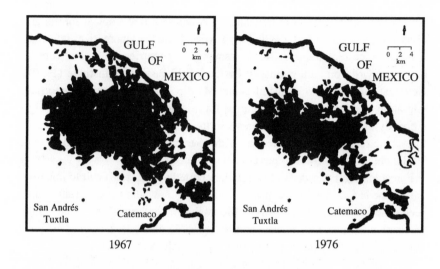

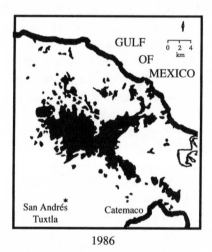

Figure 9.1. Change in forest distribution over time in the Tuxtla Mountain region of southern Veracruz, Mexico. Forested area is indicated in black. Adapted from Dirzo and Garcia (1992).

The Wood Thrush is not the only forest-dependent species of Nearctic migrant in the Tuxtlas. Seventeen species commonly winter in Tuxtla rain forest, and another 33 species occur as transients (Rappole et al. n.d.). Nor is the intensity of forest loss in the Tuxtlas unique. Many other parts of Middle America have undergone similar habitat alteration (Sader and Joyce 1988; Sader et al. 1991; Powell et al. 1992).

Figure 9.2. Former rain forest site recently cleared for pasture in the Tuxtla Mountains of southern Veracruz, Mexico.

Destruction of tropical forest has a number of wide-ranging effects on the environment that are already well documented: erosion (Duran Bernales 1970; Weyl 1972; Martinez 1973; Farnsworth and Golley 1974; Fuentes Godo 1974; Lopez Saucedo 1975; Margolis et al. 1975; Laviada 1976; Sosa Ferreyro 1976; Koolhaus 1977; Zuvekas 1978; De Ploey and Cruz 1979; Lovejoy 1981), loss of soil fertility (Farnsworth and Golley 1974; Salas and Folster 1976), reservoir siltification (Hunt 1976, 1977), and possible disruption of the earth's carbon and hydrological cycles (Lovejoy 1981). To this list can now be added the extensive negative impacts on biodiversity, including Nearctic migrants.

The economic and social returns from forest removal are questionable, even if the enormous environmental costs are ignored (Gomez-Pompa et al. 1972; Goodland and Irwin 1974; Briinig 1977; Davis 1977). Myers (1980b) summarized a number of instances in which governments have found that forest development projects, unless carefully controlled, can result in rapid loss of the forest resources and desertification of the cleared land within a few years and with little economic benefit to the nation. Despite these facts, corporate economic interests and rapid population growth continue to push governments into extensive forest destruction programs.

POLLUTION

Rappole et al. (1983) provided a bibliographic summary of 68 references on the topic of pollution in the tropical environment, indicating that contaminants are a very serious problem in the Neotropics and that many of the people who live there are aware of the problem. The well-documented experiences of the United States and other developed nations serve as stark examples of the hidden costs of pollutants to human health, livestock, crops, and, incidentally, biodiversity.

Although many governments and citizens of Neotropical countries are aware of at least some of the costs of pollution, they are forced to subject the effects of pollutants to a cost-benefit analysis. After all, in a number of tropical regions people still suffer and die from arthropod-borne diseases that can be controlled by pesticides, and a bushel of corn saved from insect pests can mean the difference between starvation and survival for some families. Given these kinds of difficult considerations, it is not surprising that use of persistent pesticides and production of industrial and human waste pollutants remain very high in developing nations.

The extent of pollutant effects on tropical biodiversity in general and on migratory birds in particular is virtually unknown. The few studies that have been done are not encouraging. A study of rice fields treated with the highly toxic and persistent pesticides endrin and NaPCP (sodium polychlorinated polyphenol) in Suriname revealed that individuals of 33 species of birds in the vicinity were suffering from pesticide poisoning. Twenty of these species, mostly ciconiiforms and falconiforms, have Nearctic migrant populations (Vermeer et al. 1974). Albuquerque (1978) examined Peregrine Falcons in Brazil and found that North American races of this species were exposed to high pesticide levels during the time spent in south coastal Brazil, though no measurements of pesticide levels were provided. Work by Schlatter (1976) in Chile and Gochfeld (1980) in the Humboldt Current off Peru further indicate that migratory waterbirds are likely to be exposed to high pesticide levels in some areas of the Neotropics.

The studies of pesticide levels in American Kestrels in New York State by Lincer and Sherburne (1974) illustrate the danger to Nearctic migrants of toxins accumulated during migration or on the wintering ground. They found that kestrels laid eggs with thinned shells and high pesticide residues, even though local prey had very low pesticide levels. The researchers concluded that the pesticides accumulated in the birds' body fat from prey captured on southern wintering grounds (southern United States, Mexico, and Central America), where potential prey items show high pesticide levels.

Studies of pesticide residues in migratory waterbirds and raptors and their eggs by Henny and his colleagues (Henny et al. 1982, 1984; Henny and Herron 1989) indicate that contamination of migrants on their wintering grounds is a serious problem, particularly for predators high in the food chain. In a study of six sites in the northwestern United States (three in Oregon, two in Washington, and one in Nevada), the investigators found that all eggs analyzed of Black-crowned Night-Herons breeding at their study sites were contaminated with DDE (a breakdown product of DDT) and that contamination had a negative effect on breeding success. Examination for local contamination at colony sites yielded no indication of DDT or DDE in nonmigratory predators (e.g., largemouth bass) for four of the six sites, including the site where bird contamination was found to be highest (Ruby Lake, Nevada). The investigators concluded that the birds must be accumulating DDT/DDE residue during migration or on their wintering grounds in Middle America, where use of DDT and similar compounds to control insect pests is still common. Interestingly, however, a telemetry study of two groups of Black-crowned Night-Herons, one of which wintered primarily in central Mexico while the other wintered principally in Arizona and southern California, revealed that the group wintering in the southwestern United States had higher levels of DDE contamination (Henny and Blus 1986). Clark and Krynitsky (1983) reported on continuing problems with banned DDT in New Mexico and Arizona. Studies of other waterbird species have raised concerns that this region of the United States remains high in both DDE and PCB contaminants, which continue to enter the food chain and cause ecological damage wherever the migratory birds that winter there travel to breed (Henny et al. 1985).

In a similar study, Henny et al. (1982) analyzed blood of migrating Peregrine Falcons during fall and spring migration. The researchers found that pesticide levels in the blood of first-year birds were extremely low during the first migration southward to wintering areas in Central America and South America. In contrast, second-year birds that were captured during northward migration in the spring had significantly higher levels of pesticide contaminants in their blood.

Determination of the magnitude of the pollutant problem for migrants should be a high research priority, not only because of its relevance for the survival of migrant populations but also because birds function as bioindicators of the general seriousness of the environmental pollution of an area. The U.S. Fish and Wildlife Service, the Canadian Fish and Wildlife Service, and the World Wildlife Fund are currently funding some research projects related to the problem of migrant contamination. More work along these lines is certainly needed.

DIRECT EXPLOITATION

Economic and cultural factors dictate the seriousness of direct forms of migratory bird exploitation. Where a tradition for hunting or a market for the products exists, there is a grave danger that the resource will be quickly exhausted. This problem occurred during the late 1800s and early 1900s in the eastern United States when hunting and a market for plumes caused the extermination of many migratory game bird populations (Eaton 1910; Bull 1974). Fortunately, neither the market nor the traditions of most of the countries of the Neotropical region have placed significant hunting pressures on migrants to date. As Saunders (1952) reported for Mexico: "Because of the much lower wage and very small average income, sport shooting can be afforded by few. For those natives who hunt game to supply needed food, ammunition is too expensive to be used on ducks where there are deer, wild pigs, and turkeys to be killed with the same ammunition."

Direct exploitation of migratory birds is not presently a conservation problem of the magnitude of habitat loss or pollution in the Neotropics. Birds are removed from wild populations for food, sport, medical research, falconry, pets, zoos, and museum collections. Yet in most cases, these activities have had little observable effect on populations. Use of migratory birds and other segments of the flora and fauna is controlled by a number of local game laws and international agreements (see section by Byron Swift in Rappole et al. 1983). Illicit trade for some species like the Peregrine Falcon is an international conservation problem, but most migrants do not fall into this category. Three principal groups of migrants have the potential to be affected by direct exploitation, particularly as their populations decline: migratory waterfowl, migratory upland game birds (mainly doves, such as Mourning Dove, White-winged Dove, and White-crowned Pigeon), and a few species of songbirds prized as pets because of the beauty of their song (e.g., Wood Thrush) or appearance (e.g., Painted Bunting) (Rappole and Ramos 1985).

HUMAN POPULATION GROWTH

Of all the threats to Nearctic migrants in the Neotropics, the most serious is that of human population growth. This single factor is the ultimate cause underlying the complex myriad of proximate factors that threaten tropical biodiversity in general and avian migrants in particular. Forest destruction, pesticides, wetland drainage, hunting pressure, industrial wastes—all of these

Table 9.2

Human population data for selected countries of the Western Hemisphere

Country	Population estimate for 1990 (millions)	Annual rate of natural increase (%)	Projected population by the year 2025 (millions)
Argentina	32.32	1.3	45.51
Belize	0.19	2.4	0.31
Bolivia	7.31	2.8	18.29
Brazil	150.37	2.1	245.81
Canada	26.52	0.9	31.92
Chile	13.17	1.7	19.77
Colombia	32.98	2.0	54.20
Costa Rica	3.02	2.6	5.25
Cuba	10.61	1.0	12.99
Dominican Republic	7.17	2.2	11.45
Ecuador	10.59	2.6	19.92
El Salvador	5.25	1.9	11.30
Guatemala	9.20	2.9	21.67
Guyana	0.80	0.2	1.16
Haiti	6.51	2.1	13.23
Honduras	5.14	3.2	11.51
Mexico	88.60	2.2	150.06
Nicaragua	3.87	3.4	9.22
Panama	2.42	2.1	3.86
Paraguay	4.28	2.9	9.18
Peru	21.55	2.1	37.35
Suriname	0.42	1.9	0.66
Trinidad and Tobago	1.28	1.7	1.98
United States	249.22	0.8	299.88
Uruguay	3.09	0.6	3.69
Venezuela	19.74	2.6	38.00

Source: Based on data from the World Resources Institute (1992).

problems are by-products of the overwhelming population growth that has occurred and is presently occurring in the Neotropics. People must have food, clothing, places to live, and a decent standard of living. With current population densities and rates of natural increase (Table 9.2), the costs to the environment for these simple needs will be devastating.

Table 9.3
Species of migrants most likely to decline in the near future

Species	Most vulnerable period[a]	Status[b]	Factors causing vulnerability[c]
Egretta caerulea	R	4	P, R, H, C, D
E. rufescens	W	3	P, R, H, C, D
Plegadis falcinellus	W	3	P, R, H, C, D
Falco peregrinus	W	4	P, C
Rallus elegans	W	3	R, H, C
Grus canadensis	M	4	R, H, C
G. americana	M	1	P, R, M
Pluvialis dominica	M	3	R, H, C
Charadrius melodus	W	4	R, H, C
Numenius borealis	M	1	P
Calidris canutus	M	3	P, M
C. himantopus	M	3	P, M
Tryngites subruficollis	W	2	P, M, R, D
Phalaropus lobatus	M	3	M
Sterna nilotica	R	2	D, P
S. dougallii	R, W	2	P, D, R, H
S. hirundo	R	3	H, D
S. antillarum	R	2	H, D
S. fuscata	R	2	P, D, R, H
Columba leucocephala	R	3	R, D
Coccyzus erythropthalmus	W	3	H
Caprimulgus carolinensis	W	3	H
Empidonax flaviventris	W	3	H
E. virescens	W	3	H
Catharus mustelinus	W	3	H
C. minimus	W	3	H
C. fuscescens	W	3	H
Vireo atricapillus	W	3	P, R
V. philadelphicus	W	3	P, R, H
V. flavoviridis	W	3	H
Vermivora bachmanii	W	1	P, R, H
V. pinus	W	3	H, R
V. chrysoptera	W	3	H
V. crissalis	W	3	P, R
V. luciae	W	3	P, R
Dendroica pensylvanica	W	3	H, R
D. tigrina	W	3	R, H

Species	Most vulnerable period[a]	Status[b]	Factors causing vulnerability[c]
D. caerulescens	W	3	R, H
D. townsendi	W	3	H
D. virens	W	3	H
D. chrysoparia	W	1	P, R, H
D. kirtlandii	W	1	H
D. castanea	W	3	R, H
D. cerulea	W	3	H
Protonotaria citrea	W	3	H
Helmitheros vermivorus	W	3	H
Limnothlypis swainsonii	W	3	H, R
Seiurus aurocapillus	W	3	H
S. motacilla	W	3	H
Oporornis formosus	W	3	H
O. agilis	W	3	P, H
Wilsonia citrina	W	3	H

[a] R = reproduction; M = migration; W = wintering.

[b] 1 = almost extinct; 2 = high probability of showing serious declines in the next decade; 3 = high probability of showing declines in the next decade; 4 = vulnerable but may or may not show declines.

[c] P = small population size; C = exposure to high levels of contaminants; D = direct disturbance by humans, such as hunting, disruption of breeding colonies, and so on; H = habitat loss; R = restricted range; M = undetermined factors at migratory stopover points.

MIGRANT VULNERABILITY

When rates of population growth, habitat destruction, and pollution are considered in light of where the highest concentrations of wintering and transient migrants occur, it is clear that the West Indies, Central America, and Mexico are the most critical regions for migrant conservation efforts at present. Species with small populations that concentrate in these regions during migration or winter are especially vulnerable to severe declines. Table 9.3 lists those species whose vulnerability appears to be highest based on the size of their populations, location of wintering and transient stopover points, and needs relative to current patterns of habitat destruction. Research is urgently needed on these and comparable species to document carrying capacities of wintering grounds and stopover points in primary and disturbed habitats, evidence of breeding population declines, and basic nonbreeding ecological data.

CONSERVATION ATTITUDES TOWARD WILDLIFE

Until very recently, most people in the Neotropics were pitted against their environment in a desperate struggle to survive. Many people in these countries are still in that position. As Tosi (1963:340) noted, "In all of the countries [of Latin America], without exception, large portions of the population live as they lived 100 years ago or more; rural people, agrarian and pre-industrial, ignorant and poor, dependent on the primitive technology and economy of the barter system at a subsistence or semisubsistence level" (translated from Spanish).

Under these circumstances, forests stand in the way of farmers trying to produce enough food to live and rivers are just conduits for travel, water supply, and wastes. Standards of living and education must change for these attitudes to change. Development of a strong conservation ethic requires political and economic stability of a sort that is just beginning to develop in many parts of the Neotropics.

Tosi (1963) discussed the terrible problems faced by peasants trying to make a living in Colombia. The good lands are already taken, so many people go to the cities, where they live in abject poverty. Others move to the mountains, where their efforts to clear land and raise crops end in disaster for themselves and for the land, whose soils and hydrologic cycles are destroyed once the covering forests are destroyed. Many of these people then leave the mountains "to demoralize the more productive rural areas with a cruel and insensate brigandage." Tosi (1963:40) continued, "And this social and economic instability is increasing in rural areas throughout all of Latin America" (translated from Spanish).

Nevertheless, efforts have been made in nearly all countries of the Western Hemisphere to preserve lands as parks or reserves (Table 9.4). Bibliographic information on these efforts is summarized in Rappole et al. (1983). Costa Rica has set aside roughly 12% of its territory, and Panama has protected more than 17% of its total area. However, Colombia's conservation efforts are more typical of efforts of the larger developing countries. Colombia passed its first wildlife protection laws in 1919, passed its first hunting laws in 1941, and created its first biological reserve in 1948 (Hunsaker 1972). Today it has roughly 8% of its total area protected as parks or reserves, most of which were set aside within the past decade. For years, actions protecting any of Colombia's lands were thanks to the efforts of a small but very active group of conservationists. Hunsaker (1972:443) noted, "As in all developing countries with rapidly expanding economies and populations, the greatest problems for conservation are the destruction of habitat by settlers and commercial (or personal) use of endangered species for economic gain or food." Karr (1978:131) made a similar point in discussing conservation problems in the Panama Canal Zone: "The major reasons for these losses [of species] are habitat destruction, hunting, pet trade, and continuing fragmentation of habitat."

Economic and social stability are necessary prerequisites to long-term, far-sighted policies of resource use. In many Western Hemisphere nations, the atmosphere for such planning has not existed, at least until quite recently, and the process of desertification continues unabated (Andrade and Payan 1973; Goodland and Irwin 1974; Andrade 1975; Gregersen and Contreras 1975; Peralta 1977; Lovejoy 1981).

RECOMMENDATIONS

Preservation of our migrant avifauna will require action in three major areas: national and international policy, research, and management. The listings under each major topic are given in order of priority.

Policy

U.S. Government Efforts
Part of the threat to our Nearctic migrant avifauna derives from the policies of other countries. However, the United States includes within its territorial boundaries several tropical, subtropical, and temperate regions that serve as key stopover and/or wintering areas for Nearctic migrants—Hawaii, Puerto Rico, the U.S. Virgin Islands, Florida, the Mississippi Flyway, the California coast, and the Gulf coastal plain of Texas, Louisiana, Mississippi, and Alabama, to mention only a few of the most important and well-known areas. Before we can ask other countries to cooperate with us in preserving a common resource, we must be able to demonstrate that we are already doing our best to enhance and preserve these avian resources. Thus, our national environmental policy should reflect our awareness and concern for the problem of avian population decline, setting an example for other nations to follow. This policy should include political efforts to help migrants, funds for cooperative research among the United States and other Western Hemisphere nations, and management directives for U.S. Fish and Wildlife Service personnel.

Emphasis on Environmental Impact Statements
Environmental impact statements are currently a prerequisite for many development projects. These statements should be required to include seasonal

Table 9.4

Categories of protected areas in selected countries of the Western Hemisphere

Country	Areas totally protected (km²)[a]	Areas partially protected (km²)[b]	Coastal areas (km²)[c]	Biosphere reserves (km²)	World heritage sites	Percentage of national land area protected
Argentina	22,690	103,700	14,990	24,100	2	4.6
Belize	40	700	—	0	0	3.2
Bolivia	26,780	40,970	NA	4,350	0	6.2
Brazil	139,060	66,190	20,320	0	1	2.4
Chile	83,780	52,710	100,500	24,070	0	18.0
Colombia	92,540	470	6,150	25,140	0	8.2
Costa Rica	4,760	1,300	1,940	7,290	1	11.9
Cuba	4,430	2,720	2,270	3,240	0	6.4
Dominican Republic	70	0	2,700	0	0	0.1
Ecuador	26,570	80,280	89,750	14,460	2	37.7

El Salvador	210	60	0	0	0	1.2
Guatemala	690	190	130	0	1	0.8
Guyana	120	0	0	0	0	0.1
Haiti	80	0	0	0	0	0.3
Honduras	5,890	1,210	3,500	5,000	1	6.3
Mexico	22,230	71,970	11,190	12,880	1	4.8
Nicaragua	270	160	40	0	0	0.3
Panama	11,950	1,310	8,980	5,970	2	17.2
Paraguay	11,570	290	NA	0	0	2.9
Peru	25,310	29,870	7,100	25,070	4	4.3
Suriname	5,530	2,100	1,280	0	0	4.7
Trinidad and Tobago	0	150	30	0	0	3.0
Uruguay	150	160	30	2,000	0	0.2
Venezuela	88,690	113,960	7,040	0	0	22.2

Source: Based on data from the World Resources Institute (1992).

[a]International Union for Conservation of Nature (IUCN) categories I–III.

[b]IUCN categories IV and V.

[c]NA = not applicable; — = not available.

assessments so that the effects of proposed development on transients and winter residents as well as on breeding residents can be properly evaluated.

Government and Corporate Environmental Responsibility

Corporations should be required to follow explicit environmental guidelines, just as some are now required by their stockholders to follow certain racial policies. U.S. corporations and government agencies play important roles in providing the necessary capital and expertise for economic development in many countries of the Neotropics. Economic, social, and environmental exploitation can have devastating, long-term effects on the developing nation and on its relations with the United States. Environmental exploitation, as pointed out in this treatise, can also have long-term effects on U.S. avifauna.

Dissemination of Research Results

A great deal of research has already been done on the various effects of different kinds of manipulation in different environments in the Tropics. Billions of investment dollars have been wasted on exploitation policies that had long-term negative results that were predictable based on knowledge available at the time of investment (Wells and Brandon 1992). These research data could and should be made available, along with sound advice in intelligible form, to U.S. and foreign investors in developing Neotropical countries. Wise use of resources will benefit both the investor and the sovereign nation in the long run. These data could be made available through U.S. government agencies, the Organization of American States, or the United Nations.

Conservation Organization Efforts

International conservation organizations should be made aware of the magnitude and the insidious nature of the potential and actual decline in Nearctic migrant and resident avifauna due to habitat destruction and other factors in the Tropics. These organizations could cooperate with governments in helping to inculcate a conservation ethic within a country, publicize alternative policies of habitat management, and assist local conservation organizations in selecting, obtaining, and maintaining critical habitat.

International Agreements

The U.S. government should seek further multilateral commitments to the protection of migratory birds as an international resource shared with all other Western Hemisphere nations. Such agreements should emphasize the need for habitat management policies that minimize effects on avian resources and that provide incentives for protecting critical habitats.

Habitat Assessment Program

A program should be established to assess the amounts of various habitats in the Neotropics. The information accumulated in this program should be disseminated to Western Hemisphere national governments and to international conservation organizations so that they can target critical areas.

Wildlife Personnel Training Program

A program should be established to train wildlife personnel from the United States and other interested nations in the management of migratory bird resources and other aspects of biodiversity.

Research

Cooperative Research Projects

The highest priority for research on Nearctic migrants is to emphasize the need for joint projects involving scientists from the United States, Canada, and the various nations of the Neotropics. It is essential that the international nature of the problem of avian resource decline be recognized from within the various countries affected. As was emphasized in the body of this book, decline in Nearctic migrants is only a symptom of the devastating alteration of environments taking place throughout the hemisphere. The resident component of the avifauna is disappearing as well as all of the other flora and fauna of the affected habitats.

Effects of Habitat Alteration

The effects of habitat alteration at stopover points and wintering areas should be determined through research.

Migrant Carrying Capacity by Habitat

The amounts and locations of major habitat types in the Neotropics should be assessed, as well as the carrying capacities of these habitats with regard to migratory bird species. These studies should involve determination of migrant species' survivorship by habitat.

Migrant Ecology and Distribution

Work should be initiated on the winter ecology (behavior, home range, and resource needs) and distribution of species most vulnerable to disturbance (see Table 9.3).

Pollutant Problems

The nature, severity, and extent of pollution in the Neotropics and its effects on migratory birds need to be determined.

Transient Ecology

Baseline data on the ecology of transients at stopover points are needed. Presently, little is known regarding timing of migration, route selection, or accumulation of resources in different habitat types along the route. Are stopover habitats critical for migrant conservation, as well as habitats in breeding and wintering areas? Shorebird research would indicate that they are, but little is known about the needs of songbirds in transit.

Migrant Population Levels

The North American Breeding Bird Survey should be continued. Despite its problems, these efforts represent the best long-term data set on continent-wide populations that we have. Furthermore, U.S. Fish and Wildlife Service statisticians have worked hard to understand the limitations of the method and improve the quality of the data (Sauer et al. 1994). In addition to Breeding Bird Survey data, intensive multiyear research should continue in undisturbed temperate region habitats to monitor changes in populations. Also, efforts should be made to assess changes in numbers of birds in transit.

Conservation Laws and Enforcement

Information on the conservation statutes and enforcement climate in the various countries in the Western Hemisphere should be researched and developed.

Land-Use Alternatives

Studies to determine methods of land use that are compatible with the environment should continue to receive funding. Plans for the dissemination of the results of such projects in understandable form to land users and investors should be included as part of the initial research proposal.

Nonpasserine Research

Efforts should be made to initiate non-breeding-season studies on nonpasserines. These groups have been largely neglected in previous ecological research in the Neotropics.

Management

Personnel Training

Management efforts at enhancing populations of both game and nongame species should be more evenly divided between breeding and nonbreeding seasons, rather than concentrated on the former, as has been done in the past. Persons involved at the primary levels of management of national parks, seashores, monuments, wildlife refuges, forests, and other federal lands should be made aware of this need.

Seasonal Population Inventories

Population inventories of migratory birds using federal lands should be conducted seasonally to determine land use by these species. These inventories should form the basis for determination of designation as critical habitat.

Preservation of Critical Habitat

Until more specific information is available, management for Nearctic migrants should consist largely of obtaining and preserving habitats. Efforts should be made to identify, obtain, and protect as management areas critical stopover and wintering sites, particularly for those species whose populations tend to concentrate at a few such sites.

Management of U.S. Federal Lands in the Neotropics

Special attention should be directed toward managing federal lands in Hawaii, the U.S. Virgin Islands, and Puerto Rico for preserving habitat for Nearctic migrants.

Conservation Education Programs

There is a need to make taxpaying citizens aware of the nature and breadth of the problems facing migratory birds. This development of public awareness could be done through literature and programs presented by public relations personnel at national parks, seashores, wildlife refuges, monuments, forests, and other areas where management personnel come in contact with the public.

Reintroduction of Exterminated Populations

Along the U.S. Atlantic coast, efforts should be made to reestablish as transient or winter resident populations some of the many species that were common in this region prior to settlement but have since been all but eradicated there.

Banding and Marking Programs

The United States should assist other nations of the Western Hemisphere in the development of international banding and marking programs to assess intertropical and intratropical movements. The principal form of assistance should be in serving as a central repository for data on both migrant and resident species, particularly because what constitutes a "migrant" versus a "resident" is fuzzy at best. Also, the U.S. banding program should provide bands to banders in nations that wish to participate in an international banding program.

SUMMARY

The emphasis of this final chapter has been on conservation of migratory birds from a tropical perspective. This emphasis should not be interpreted as a slight to the importance of the breeding area. A number of excellent reviews have been done elsewhere on the importance of breeding areas to migratory species (Askins et al. 1990; Hagan and Johnston 1992). In this treatise, we have examined a small part (some would say an insignificant part) of the environment—the migratory birds that breed in North America and winter in the Tropics.

Of what significance can birds be when war and famine threaten large segments of the world's human population? Birds, especially migrants, reflect the health of the environment, the interconnection between environmental deterioration in one area and its direct effects on another area thousands of kilometers away. Other examples of the possible consequences of grand-scale environmental manipulation include acid rain, the greenhouse effect, and disruption of the carbon cycle. Each of these symptoms has a number of frightening but as yet unproven set of possible repercussions. But we know something about the birds. We know about their ecology and what happens to them when their wintering habitats are destroyed. Natural balances are being tipped in the oldest terrestrial and aquatic ecosystems on earth. Migrants constitute a bond between these and other systems, a vivid illustration of how the tipping of the scales in the Tropics can cause the disruption of temperate zone ecosystems. Birds are a small part of the system—but they are a part about which we have some understanding regarding the consequences of our actions. We know that if tropical environments are destroyed, then the birds associated with them, migrants as well as residents, will be destroyed. Their fate should serve as a clear and unequivocal warning that national boundaries are not recognized by the forces of disruption of natural systems.

APPENDIX 1: LIST OF NEARCTIC–NEOTROPICAL MIGRANTS

The following list of Nearctic–Neotropical migrants contains 338 species of birds, all or part of whose populations breed north of the tropic of Cancer and winter south of that line. The list is presented in taxonomic order following the American Ornithologists' Union's *Check-list of North American Birds,* sixth edition (1983), and supplements 35 through 39. The English names follow the same authority.

The main guide used in formulating the Spanish names was the Spanish translation of Peterson and Chalif's *Mexican Birds* (1989), by Ramos and Castillo. I followed the recommendations contained in this excellent summary, except where names were based on Indian dialects or local usage apparently unfamiliar to workers in other parts of Latin America.

If a species has resident breeding populations somewhere in the Tropics, its scientific name is followed by two asterisks (**) in the list. If no members of the species are known to be tropical residents, but one or more congeners breed in the Tropics, the species is marked with a single asterisk (*). If there is some question as to whether a conspecific or congeneric population breeds in the Tropics, the scientific name is followed by either two question marks (??) or one question mark (?), respectively.

The close relationships that exist between Nearctic and Neotropical avifaunas are reflected by the taxonomic debates concerning whether or not members of populations should be considered as members of the same species, even if one population breeds in the Tropics and the other breeds in the temperate zone. In most cases, where one or more taxonomic authorities recommend that such separate populations be considered conspecific, I have listed them with asterisks, even though the current AOU *Check-list* may list them as separate species. As Lack (1956) stated, "Where there is doubt concerning closely related allopatric forms, it is in general better to classify them as subspecies, not species. . . . [T]his has the advantage of indicating affinity."

The scientific names of the species affected by this procedure are followed by a superscript number where I am in disagreement with the AOU *Check-list* (1983) and supplements. By referring to the corresponding notes at the end of this appendix, the reader can find the authorities followed in making the taxonomic decision as well as the semispecies that have been lumped together as populations of the same species within a particular group. The purpose of this treatment is to emphasize the ecological and evolutionary affinities of two supposedly separate groups, Nearctic migrants and tropical residents, not to make a taxonomic statement.

Scientific name	English name	Spanish name
PODICIPEDIDAE		
*Podilymbus podiceps***	Pied-billed Grebe	Zambullidor Piquigrueso
*Podiceps nigricollis***	Eared Grebe	Zambullidor Mediano
*Aechmophorus occidentalis***	Western Grebe	Zambullidor Achichilique
*A. clarkii***	Clark's Grebe	Zambullidor de Clark
PELECANIDAE		
*Pelecanus erythrorhynchos**	American White Pelican	Pelícano Blanco
*P. occidentalis***	Brown Pelican	Pelícano Pardo
PHALACROCORACIDAE		
*Phalacrocorax auritus***	Double-crested Cormorant	Cormorán Orejudo
*P. olivaceus***	Olivaceous Cormorant	Cormorán Oliváceo
ANHINGIDAE		
*Anhinga anhinga***	Anhinga	Anhinga Americana
ARDEIDAE		
*Botaurus lentiginosus**	American Bittern	Garza Norteña de Tular
*Ixobrychus exilis***	Least Bittern	Garcita de Tular
*Ardea herodias***	Great Blue Heron	Garzón Cenizo
*Casmerodius albus***	Great Egret	Garzón Blanco
*Egretta thula***	Snowy Egret	Garza Dedos Dorados
*E. caerulea***	Little Blue Heron	Garza Azul
*E. tricolor***	Tricolored Heron	Garza Ventriblanca
*E. rufescens***	Reddish Egret	Garza Piquirrosa
*Bubulcus ibis***	Cattle Egret	Garza Ganadera
*Butorides striatus***[1]	Green-backed Heron	Garcita Oscura
*Nycticorax nycticorax***	Black-crowned Night-Heron	Garza Nocturna Coroninegra
*Nyctanassa violacea***	Yellow-crowned Night-Heron	Garza Nocturna Coroniclara
THRESKIORNITHIDAE		
*Eudocimus albus***	White Ibis	Ibis Blanco
*Plegadis falcinellus***	Glossy Ibis	Ibis Negro
*P. chihi***	White-faced Ibis	Ibis Cara Blanca
*Ajaia ajaja***	Roseate Spoonbill	Ibis Espátula
CICONIIDAE		
*Mycteria americana***	Wood Stork	Cigüeña Americana
ANATIDAE		
*Dendrocygna bicolor***	Fulvous Whistling-Duck	Pato Pijije Alioscuro
*D. autumnalis***	Black-bellied Whistling-Duck	Pato Pijije Aliblanco
Anser albifrons	Greater White-fronted Goose	Ganso Manchado
Chen caerulescens	Snow Goose	Ganso Cerúleo (Blanco)
*Aix sponsa***	Wood Duck	Pato Arcoiris
*Anas crecca**	Green-winged Teal	Cerceta Alioscura
*A. platyrhynchos***[2]	Mallard	Pato de Collar
*A. acuta**	Northern Pintail	Pato Golondrino
A. discors??	Blue-winged Teal	Cerceta Aliazul Clara

Scientific name	English name	Spanish name
*A. cyanoptera***	Cinnamon Teal	Cerceta Aliazul Café
*A. clypeata**	Northern Shoveler	Pato Cucharón
*A. strepera**	Gadwall	Pato Friso
*A. americana**	American Wigeon	Pato Chalcuán
*Aythya valisineria**	Canvasback	Pato Borrado
*A. americana***	Redhead	Pato Cabecirrojo
*A. collaris**	Ring-necked Duck	Pato Piquianillado
*A. affinis**	Lesser Scaup	Pato Boludo Menor
Lophodytes cucullatus	Hooded Merganser	Pato Mergo Copetón
Mergus serrator	Red-breasted Merganser	Pato Mergo Pechicastaño
*Oxyura jamaicensis***[3]	Ruddy Duck	Pato Rojizo Alioscuro

CATHARTIDAE

*Coragyps atratus***	Black Vulture	Carroñero Común
*Cathartes aura***	Turkey Vulture	Aura Común

ACCIPITRIDAE

*Pandion haliaetus***	Osprey	Aguila Pescadora
*Elanoides forficatus***	American Swallow-tailed Kite	Milano Tijereta
*Ictinia mississippiensis***[4]	Mississippi Kite	Milano Migratorio
*Circus cyaneus***[5]	Northern Harrier	Aguililla Rastrera
*Accipiter striatus***[6]	Sharp-shinned Hawk	Gavilán Pechirrufo Menor
*A. cooperii***	Cooper's Hawk	Gavilán Pechirrufo Mayor
*Buteogallus anthracinus***	Common Black-Hawk	Aguililla Negra Menor
*Buteo platypterus***	Broad-winged Hawk	Aguililla Migratoria Menor
*B. swainsoni**	Swainson's Hawk	Aguililla Migratoria Mayor
*B. jamaicensis***	Red-tailed Hawk	Aguililla Colirrufa
*B. regalis**	Ferruginous Hawk	Aguililla Real

FALCONIDAE

*Falco sparverius***	American Kestrel	Halcón Cernícalo
*F. columbarius**	Merlin	Halcón Esmerejón
*F. peregrinus***	Peregrine Falcon	Halcón Peregrino
*F. mexicanus**	Prairie Falcon	Halcón Pálido

RALLIDAE

*Laterallus jamaicensis***	Black Rail	Ralito Negruzco
*Rallus elegans***	King Rail	Ralón Barrado Rojizo
*R. limicola***	Virginia Rail	Ralo Barrado Rojizo
*Porzana carolina***	Sora	Ralo Barrado Grisáceo
*Porphyrula martinica***	Purple Gallinule	Gallareta Morada
*Gallinula chloropus***	Common Moorhen	Gallareta Frentirroja
*Fulica americana***	American Coot	Gallareta Americana

GRUIDAE

*Grus canadensis***	Sandhill Crane	Grulla Gris
*G. americana**	Whooping Crane	Grulla Blanca

CHARADRIIDAE

Pluvialis squatarola	Black-bellied Plover	Chorlo Axilinegro

Scientific name	English name	Spanish name
P. dominica	Lesser Golden-Plover	Chorlo Axiliclaro
*Charadrius alexandrinus***	Snowy Plover	Chorlito Alejandrino
*C. wilsonia***	Wilson's Plover	Chorlito Piquigrueso
*C. semipalmatus**	Semipalmated Plover	Chorlito Semipalmeado
*C. melodus**	Piping Plover	Chorlito Melódico
*C. vociferus***	Killdeer	Chorlito Tildío
*C. montanus**	Mountain Plover	Chorlito Llanero

HAEMATOPODIDAE

*Haematopus palliatus***	American Oystercatcher	Ostrero Blanquinegro

RECURVIROSTRIDAE

*Himantopus mexicanus***	Black-necked Stilt	Avoceta Piquirrecta
*Recurvirostra americana***	American Avocet	Avoceta Piquicurva

SCOLOPACIDAE

Tringa melanoleuca	Greater Yellowlegs	Patamarilla Mayor
T. flavipes	Lesser Yellowlegs	Patamarilla Menor
T. solitaria	Solitary Sandpiper	Playero Charquero
*Catoptrophorus semipalmatus***	Willet	Playero Pihuihui
Heteroscelus incanus	Wandering Tattler	Playero Sencillo
Actitis macularia	Spotted Sandpiper	Playerito Alzacolita
Bartramia longicauda	Upland Sandpiper	Zarapito Ganga
Numenius borealis	Eskimo Curlew	Zarapito Boreal
N. phaeopus	Whimbrel	Zarapito Cabecirrayado
N. americanus	Long-billed Curlew	Zarapito Piquilargo
Limosa haemastica	Hudsonian Godwit	Limosa Ornamentada
L. fedoa	Marbled Godwit	Limosa Canela
Arenaria interpres	Ruddy Turnstone	Vuelvepiedras Común
Aphriza virgata	Surfbird	Playero Roquero
Calidris canutus	Red Knot	Playero Piquicorto
C. alba	Sanderling	Playerito Correlón
C. pusilla	Semipalmated Sandpiper	Playerito Semipalmeado
C. mauri	Western Sandpiper	Playerito Occidental
C. minutilla	Least Sandpiper	Playerito Mínimo
C. fuscicollis	White-rumped Sandpiper	Playerito de Rabadilla Blanca
C. bairdii	Baird's Sandpiper	Playerito de Baird
C. melanotos	Pectoral Sandpiper	Playero Pechirrayado
C. himantopus	Stilt Sandpiper	Playero Zancón
Tryngites subruficollis	Buff-breasted Sandpiper	Playerito Pradero
Limnodromus griseus	Short-billed Dowitcher	Costurero Marino
L. scolopaceus	Long-billed Dowitcher	Costurero de Agua Dulce
*Gallinago gallinago***[7]	Common Snipe	Agachona Común
Phalaropus tricolor	Wilson's Phalarope	Falaropo Piquilargo
P. lobatus	Red-necked Phalarope	Falaropo Piquifino
P. fulicaria	Red Phalarope	Falaropo Piquigrueso

LARIDAE

*Larus atricilla***	Laughing Gull	Gaviota Atricila
*L. pipixcan**	Franklin's Gull	Gaviota Apipizca

Scientific name	English name	Spanish name
L. philadelphia*	Bonaparte's Gull	Gaviota Menor
L. delawarensis*	Ring-billed Gull	Gaviota de Delaware
L. californicus*	California Gull	Gaviota Californiana
L. argentatus*	Herring Gull	Gaviota Plateada
L. occidentalis*	Western Gull	Gaviota Occidental
L. glaucescens*	Glaucous-winged Gull	Gaviota Aliglauca
Xema sabini	Sabine's Gull	Gaviota Colihendida
Sterna nilotica**	Gull-billed Tern	Gaviotina Piquigruesa
S. caspia**	Caspian Tern	Gaviotina Piquirroja
S. maxima**	Royal Tern	Gaviotina Piquinaranja
S. elegans**	Elegant Tern	Gaviotina Elegante
S. sandvicensis**[8]	Sandwich Tern	Gaviotina de Sandwich
S. dougallii**	Roseate Tern	Gaviotina de Dougall
S. hirundo**	Common Tern	Gaviotina Común
S. forsteri*	Forster's Tern	Gaviotina de Forster
S. antillarum**	Least Tern	Gaviotina Menor
S. fuscata**	Sooty Tern	Gaviotina Dorsinegra
Chlidonias niger	Black Tern	Gaviotina Negruzca
Rynchops niger**	Black Skimmer	Rayador Americano

COLUMBIDAE

Columba leucocephala**	White-crowned Pigeon	Paloma Cabeciblanca
C. flavirostris**	Red-billed Pigeon	Paloma Morada Ventrioscura
C. fasciata**	Band-tailed Pigeon	Paloma Collareja
Zenaida asiatica**	White-winged Dove	Paloma Aliblanca
Z. macroura**	Mourning Dove	Paloma Común

CUCULIDAE

Coccyzus erythropthalmus*	Black-billed Cuckoo	Cuclillo Piquinegro
C. americanus*	Yellow-billed Cuckoo	Cuclillo Alirrojizo
C. minor**	Mangrove Cuckoo	Cuclillo Ventrisucio

STRIGIDAE

Micrathene whitneyi**	Elf Owl	Lechucilla Colicorta
Speotyto cunicularia**	Burrowing Owl	Lechucilla Zancón
Asio flammeus**	Short-eared Owl	Buho Cornicorto Llanero

CAPRIMULGIDAE

Chordeiles acutipennis**	Lesser Nighthawk	Chotacabra Halcón
C. minor**	Common Nighthawk	Chotacabra Zumbón
Phalaenoptilus nuttallii**	Common Poorwill	Tapacamino Tevíi
Caprimulgus carolinensis**	Chuck-will's-widow	Tapacamino de Paso
C. vociferus**[9]	Whip-poor-will	Tapacamino Cuerporruín

APODIDAE

Cypseloides niger**	Black Swift	Vencejo Negro
Chaetura pelagica**[10]	Chimney Swift	Vencejito de Paso
C. vauxi**	Vaux's Swift	Vencejito Alirrápido
Aeronautes saxatalis**	White-throated Swift	Vencejito Pechiblanco

Scientific name	English name	Spanish name
TROCHILIDAE		
*Cynanthus latirostris***	Broad-billed Hummingbird	Colibrí Latirrostro
*Amazilia yucatanensis***	Buff-bellied Hummingbird	Amazilia del Golfo
*A. violiceps***	Violet-crowned Hummingbird	Amazilia Occidental
*Lampornis clemenciae***	Blue-throated Hummingbird	Chupaflor Gorjiazul
*Eugenes fulgens***	Magnificent Hummingbird	Chupaflor Coronimorado
Archilochus colubris	Ruby-throated Hummingbird	Colibrí de Paso
A. alexandri	Black-chinned Hummingbird	Colibrí Gorjinegro
Calypte anna	Anna's Hummingbird	Colibrí Coronirrojo
C. costae	Costa's Hummingbird	Colibrí Coronivioleta
Stellula calliope	Calliope Hummingbird	Colibrí Gorjirrayado
*Selasphorus platycercus***	Broad-tailed Hummingbird	Colibrí Vibrador
*S. rufus**	Rufous Hummingbird	Colibrí Colicanelo Rufo
*S. sasin**	Allen's Hummingbird	Colibrí Colicanelo Sasin
TROGONIDAE		
*Trogon elegans***	Elegant Trogon	Trogón Colicobrizo
ALCEDINIDAE		
*Ceryle alcyon**	Belted Kingfisher	Martín Pescador Norteño
PICIDAE		
Sphyrapicus varius	Yellow-bellied Sapsucker	Carpintero Aliblanco Común
S. nuchalis	Red-naped Sapsucker	Carpintero Nuquirrojo
S. thyroideus	Red-breasted Sapsucker	Carpintero Aliblanco Oscuro
TYRANNIDAE		
*Camptostoma imberbe***	Northern Beardless Tyrannulet	Mosquerito Silbador
*Contopus borealis**	Olive-sided Flycatcher	Contopus de Chaleco
*C. pertinax***	Greater Pewee	Contopus José María
*C. sordidulus***	Western Wood-Pewee	Contopus Occidental
*C. virens***	Eastern Wood-Pewee	Contopus Verdoso
*Empidonax flaviventris**	Yellow-bellied Flycatcher	Empidonax Ventriamarillo
*E. virescens**	Acadian Flycatcher	Empidonax Verdoso
*E. alnorum**	Alder Flycatcher	Empidonax Alnoro
*E. traillii**	Willow Flycatcher	Empidonax de Traill
*E. minimus**	Least Flycatcher	Empidonax Mínimo
*E. hammondii**	Hammond's Flycatcher	Empidonax de Hammond
*E. oberholseri**	Dusky Flycatcher	Empidonax de Oberholser
*E. wrightii**	Gray Flycatcher	Empidonax de Wright
*E. difficilis***[11]	Pacific-slope Flycatcher	Empidonax Difícil
*E. occidentalis***[11]	Cordilleran Flycatcher	Empidonax Occidental
*E. fulvifrons***	Buff-breasted Flycatcher	Empidonax Canelo
*Sayornis phoebe**	Eastern Phoebe	Mosquero Fibí
*S. saya**	Say's Phoebe	Mosquero Llanero
*Pyrocephalus rubinus***	Vermilion Flycatcher	Mosquero Cardenalito
*Myiarchus tuberculifer***	Dusky-capped Flycatcher	Papamoscas Copetón Triste
*M. cinerascens***[12]	Ash-throated Flycatcher	Papamoscas Copetón Gorjicenizo
*M. crinitus**	Great Crested Flycatcher	Papamoscas Copetón Viajero

Scientific name	English name	Spanish name
*M. tyrannulus***	Brown-crested Flycatcher	Papamoscas Copetón Tiranillo
*Myiodynastes luteiventris***	Sulphur-bellied Flycatcher	Papamoscas Rayado Cejiblanco
*Tyrannus melancholicus***	Tropical Kingbird	Tirano Tropical Común
*T. couchii***	Couch's Kingbird	Tirano Tropical Silbador
*T. vociferans**	Cassin's Kingbird	Tirano Gritón
*T. crassirostris***	Thick-billed Kingbird	Tirano Piquigrueso
*T. verticalis**	Western Kingbird	Tirano Pálido
*T. tyrannus**	Eastern Kingbird	Tirano Dorsinegro
*T. dominicensis***	Gray Kingbird	Tirano Dominicano
*T. forficatus**	Scissor-tailed Flycatcher	Tirano Tijereta Clara

HIRUNDINIDAE

*Progne subis***[13]	Purple Martin	Golondrina Grande Negruzca
*Tachycineta bicolor**	Tree Swallow	Golondrina Canadiense
*T. thalassina***	Violet-green Swallow	Golondrina Cariblanca
*Stelgidopteryx serripennis***	Northern Rough-winged Swallow	Golondrina Gorjicafé
Riparia riparia	Bank Swallow	Golondrina Pechifajada
*Hirundo pyrrhonota***	Cliff Swallow	Golondrina Risquera
*H. fulva***	Cave Swallow	Golondrina Fulva
*H. rustica***	Barn Swallow	Golondrina Tijereta

TROGLODYTIDAE

*Troglodytes aedon***[14]	House Wren	Troglodita Continental
*Cistothorus platensis***	Sedge Wren	Troglodita de Ciénega
*C. palustris***	Marsh Wren	Troglodita Pantanera

MUSCICAPIDAE

*Regulus calendula**	Ruby-crowned Kinglet	Reyezuelo Sencillo
*Polioptila caerulea***	Blue-gray Gnatcatcher	Perlita Piis
*Sialia sialis***	Eastern Bluebird	Azulejo Gorjicanelo
S. mexicana	Western Bluebird	Azulejo Gorjiazul
*S. currucoides***	Mountain Bluebird	Azulejo Pálido
*Myadestes townsendi***	Townsend's Solitaire	Clarín Norteño
*Catharus fuscescens**	Veery	Zorzalito de Wilson
*C. minimus**	Gray-cheeked Thrush	Zorzalito Carigrís
*C. ustulatus**	Swainson's Thrush	Zorzalito de Swainson
*C. guttatus***	Hermit Thrush	Zorzalito Colirrufo
*C. mustelinus**	Wood Thrush	Zorzalito Maculado
*Turdus migratorius***	American Robin	Zorzal Pechirrojo

MIMIDAE

*Dumetella carolinensis**[15]	Gray Catbird	Mímido Gris
Oreoscoptes montanus	Sage Thrasher	Mímido Pinto

MOTACILLIDAE

*Anthus rubescens**	American Pipit	Bisbita Americano
*A. spragueii**	Sprague's Pipit	Bisbita Llanera

Scientific name	English name	Spanish name

BOMBYCILLIDAE
Bombycilla cedrorum — Cedar Waxwing — Ampelis Americano

LANIIDAE
*Lanius ludovicianus*** — Loggerhead Shrike — Verdugo Americano

VIREONIDAE

Scientific name	English name	Spanish name
*Vireo griseus***[16]	White-eyed Vireo	Vireo Ojiblanco
*V. bellii**	Bell's Vireo	Vireo de Bell
*V. atricapillus**	Black-capped Vireo	Vireo Gorrinegro
*V. vicinior** ·	Gray Vireo	Vireo Gris
*V. solitarius***	Solitary Vireo	Vireo Anteojillo
*V. flavifrons**	Yellow-throated Vireo	Vireo Pechiamarillo
*V. gilvus***[17]	Warbling Vireo	Vireo Gorjeador Norteño
*V. philadelphicus**	Philadelphia Vireo	Vireo Filadélfico
*V. olivaceus***[18]	Red-eyed Vireo	Vireo Ojirrojo Norteño
*V. flavoviridis***	Yellow-green Vireo	Vireo Ojirrojo Tropical
*V. altiloquus***	Black-whiskered Vireo	Vireo Bigotinegro

EMBERIZIDAE

Scientific name	English name	Spanish name
*Vermivora bachmanii**	Bachman's Warbler	Chipe de Bachman
*V. pinus**	Blue-winged Warbler	Chipe Aliazul
*V. chrysoptera**	Golden-winged Warbler	Chipe Alidorado
*V. peregrina**	Tennessee Warbler	Chipe Peregrino
*V. celata**	Orange-crowned Warbler	Chipe Celato
*V. ruficapilla**	Nashville Warbler	Chipe de Nashville
*V. virginiae**	Virginia's Warbler	Chipe de Virginia
*V. crissalis**	Colima Warbler	Chipe de Colima
*V. luciae**	Lucy's Warbler	Chipe de Lucy
*Parula americana***[19]	Northern Parula	Chipe Azul-Olivo Norteño
*Dendroica petechia***[20]	Yellow Warbler	Chipe Amarillo Norteño
*D. pensylvanica**	Chestnut-sided Warbler	Chipe Gorriamarillo
*D. magnolia**	Magnolia Warbler	Chipe Colifajado
*D. tigrina**	Cape May Warbler	Chipe Tigrino
*D. caerulescens**	Black-throated Blue Warbler	Chipe Azul Pizarra
*D. coronata***	Yellow-rumped Warbler	Chipe Grupidorado
*D. nigrescens**	Black-throated Gray Warbler	Chipe Negrigrís
*D. townsendi**	Townsend's Warbler	Chipe de Townsend
*D. occidentalis**	Hermit Warbler	Chipe Occidental
*D. virens**	Black-throated Green Warbler	Chipe Dorsiverde
*D. chrysoparia**	Golden-cheeked Warbler	Chipe Dorsinegro
*D. fusca**	Blackburnian Warbler	Chipe Gorjinaranja
*D. dominica***	Yellow-throated Warbler	Chipe Dominico
*D. graciae***	Grace's Warbler	Chipe Pinero Gorjiamarillo
*D. pinus***	Pine Warbler	Chipe Nororiental
*D. kirtlandii**	Kirtland's Warbler	Chipe de Kirtland
*D. discolor***[21]	Prairie Warbler	Chipe Galán
*D. palmarum**	Palm Warbler	Chipe Playero
*D. castanea**	Bay-breasted Warbler	Chipe Pechicastaño
*D. striata**	Blackpoll Warbler	Chipe Gorrinegro

Scientific name	English name	Spanish name
*D. cerulea**	Cerulean Warbler	Chipe Cerúleo
Mniotilta varia	Black-and-white Warbler	Chipe Trepador
Setophaga ruticilla	American Redstart	Pavito Migratorio
Protonotaria citrea	Prothonotary Warbler	Chipe Cabecidorado
Helmitheros vermivorus	Worm-eating Warbler	Chipe Vermívoro
Limnothlypis swainsonii	Swainson's Warbler	Chipe Coronicafé
Seiurus aurocapillus	Ovenbird	Chipe Suelero Coronado
S. noveboracensis	Northern Waterthrush	Chipe Suelero Gorjijaspeado
S. motacilla	Louisiana Waterthrush	Chipe Suelero Gorjiblanco
Oporornis formosus?	Kentucky Warbler	Chipe Cachetinegro
O. agilis?	Connecticut Warbler	Chipe Cabecigrís Ojianillado
O. philadelphia?	Mourning Warbler	Chipe Cabecigrís Filadélfico
O. tolmiei	MacGillivray's Warbler	Chipe Cabecigrís de Tolmie
*Geothlypis trichas***	Common Yellowthroat	Mascarita Norteña
Wilsonia citrina	Hooded Warbler	Chipe Encapuchado
W. pusilla	Wilson's Warbler	Chipe Coroninegro
W. canadensis	Canada Warbler	Chipe de Collar
*Cardellina rubrifrons***	Red-faced Warbler	Chipe Carirrojo
*Myioborus pictus***	Painted Redstart	Pavito Aliblanco
*Icteria virens***	Yellow-breasted Chat	Chipe Piquigrueso
*Piranga flava***	Hepatic Tanager	Tangara Roja Piquioscura
*P. rubra**	Summer Tanager	Tangara Roja Migratoria
*P. olivacea**	Scarlet Tanager	Tangara Rojinegra Migratoria
*P. ludoviciana**	Western Tanager	Tangara Aliblanca Migratoria
*Pheucticus ludovicianus**	Rose-breasted Grosbeak	Picogrueso Pechirrosa
*P. melanocephalus**	Black-headed Grosbeak	Picogrueso Pechicafé
*Guiraca caerulea***	Blue Grosbeak	Picogrueso Azul
*Passerina amoena**	Lazuli Bunting	Colorín Aliblanco
*P. cyanea**	Indigo Bunting	Colorín Azul
*P. versicolor***	Varied Bunting	Colorín Oscuro
*P. ciris**	Painted Bunting	Colorín Sietecolores
Spiza americana	Dickcissel	Espiza
*Pipilo chlorurus**	Green-tailed Towhee	Rascador Migratorio
*P. erythrophthalmus***	Rufous-sided Towhee	Rascador Pinto Oscuro
*Aimophila botterii***[22]	Botteri's Sparrow	Gorrión de Botteri
A. carpalis	Rufous-winged Sparrow	Gorrión Bigotudo Sonorense
*A. ruficeps**	Rufous-crowned Sparrow	Gorrión Bigotudo Coronirrufo
*Spizella passerina***	Chipping Sparrow	Gorrión Llanero
*S. pallida**	Clay-colored Sparrow	Gorrión Indefinido Rayado
*S. breweri**	Brewer's Sparrow	Gorrión de Brewer
*S. atrogularis**	Black-chinned Sparrow	Gorrión Indefinido Oriental
Pooecetes gramineus	Vesper Sparrow	Gorrión Zacatero Coliblanco
Chondestes grammacus	Lark Sparrow	Gorrión Arlequín
Calamospiza melanocorys	Lark Bunting	Llanero Alipálido
*Passerculus sandwichensis***	Savannah Sparrow	Gorrión Sabanero Común
*Ammodramus savannarum***	Grasshopper Sparrow	Gorrión Saltamonte
*Melospiza lincolnii**	Lincoln's Sparrow	Gorrión de Lincoln
*M. georgiana**	Swamp Sparrow	Gorrión Georgiana
*Zonotrichia leucophrys**	White-crowned Sparrow	Gorrión Gorriblanco

Scientific name	English name	Spanish name
Dolichonyx oryzivorus	Bobolink	Tordo Migratorio
*Agelaius phoeniceus***	Red-winged Blackbird	Tordo Sargento
*Sturnella magna***	Eastern Meadowlark	Pradero Tortilla-con-chile
*S. neglecta**	Western Meadowlark	Pradero Gorjeador
Xanthocephalus xanthocephalus?	Yellow-headed Blackbird	Tordo Cabeciamarillo
Euphagus cyanocephalus?	Brewer's Blackbird	Tordo Ojiclaro
*Molothrus aeneus***	Bronzed Cowbird	Tordo Ojirrojo
*M. ater***	Brown-headed Cowbird	Tordo Cabecicafé
*Icterus spurius***[23]	Orchard Oriole	Bolsero Castaño
*I. cucullatus***	Hooded Oriole	Bolsero Cuculado
I. graduacauda	Audubon's Oriole	Bolsero Capuchinegro
*I. galbula***[24]	Northern Oriole	Bolsero Norteño
*I. parisorum***	Scott's Oriole	Bolsero Parisino

FRINGILLIDAE

Carduelis psaltria	Lesser Goldfinch	Jilguero Dorsioscuro
*C. tristis**	American Goldfinch	Jilguero Canario

NOTES

[1] *Butorides striatus, B. virescens*—American Ornithologists' Union (1976)

[2] *Anas platyrhynchos, A. diazi*—Mayr and Short (1970)

[3] *Oxyura jamaicensis, O. ferruginea*—Meyer de Schauensee (1966)

[4] *Ictinia mississippiensis, I. plumbea*—Sutton (1944)

[5] *Circus cyaneus, C. cinereus*—Ridgway (1901–1950)

[6] *Accipiter striatus, A. erythrocnemius*—Mayr and Short (1970)

[7] *Gallinago gallinago, G. paraguaiae*—Meyer de Schauensee (1966)

[8] *Sterna sandvicensis, S. eurygnatha*—Junge and Voous (1955)

[9] *Caprimulgus vociferus, C. noctitherus*—Peters et al. (1931–1986)

[10] *Chaetura pelagica, C. chapmani*—Lack (1956)

[11] *Empidonax difficilis/E. occidentalis, E. flavescens*—Phillips et al. (1964)

[12] *Myiarchus cinerascens, M. nuttingi*—van Rossem (1936)

[13] *Progne subis, P. cryptoleuca, P. dominicensis, P. modesta*—Ridgway (1901–1950)

[14] *Troglodytes aedon, T. musculus, T. brunneicollis*—Meyer de Schauensee (1966)

[15] *Dumetella carolinensis, Melanoptila glabrirostris*—Paynter (1955)

[16] *Vireo griseus, V. pallens, V. crassirostris, V. gundlachii, V. modestus, V. caribaeus*—Hamilton (1958)

[17] *Vireo gilvus, V. leucophrys*—Hamilton (1958)

[18] *Vireo olivaceus, V. chivi*—Mayr and Short (1970)

[19] *Parula americana, P. pitiayumi, P. graysoni*—Mayr and Short (1970)

[20] *Dendroica petechia, D. erithachorides*—Meyer de Schauensee (1966)

[21] *Dendroica discolor, D. vitellina*—Bond (1957)

[22] *Aimophila botterii, A. petenica*—Webster (1959)

[23] *Icterus spurius, I. fuertesi*—Peters et al. (1931–1986)

[24] *Icterus galbula, I. abeillei*—Peters et al. (1931–1986)

APPENDIX 2: SYSTEMATIC RELATIONSHIPS AND HABITAT USE FOR PALEARCTIC–AFRICAN MIGRANTS

The list of Palearctic–African migrant species is based on Moreau (1972). Principal habitats are as follows: A = aquatic, F = forest, O = open (i.e., grassland, cropland, desert, savannah, scrub). Breeding habitats are based on Peterson et al. (1967). Winter habitats are based on Moreau (1972). Those species for which a congener is known to breed in the African Tropics are marked with an asterisk (*); those species for which a conspecific is known to breed in the African Tropics are marked with two asterisks (**), based on taxonomic data in Peters et al. (1931–1986).

Species	Breeding habitat	Winter habitat
ARDEIDAE		
*Botaurus stellaris***	A	A
*Ixobrychus minutus***	A	A
*Ardea cinerea***	A	A
*A. purpurea***	A	A
*Casmerodius albus***	A	A
*Egretta garzetta***	A	A
*Ardeola ralloides***	A	A
*Nycticorax nycticorax***	A	A
THRESKIORNITHIDAE		
*Geronticus eremita***	O	A,O
*Plegadis falcinellus***	A	A
*Platalea leucorodia***	A	A
CICONIIDAE		
Ciconia ciconia	A	A
C. nigra	A	A
ANATIDAE		
Tadorna ferruginea	A	A
*Anas crecca**	A	A
*A. platyrhynchos**	A	A
*A. acuta**	A	A
*A. querquedula**	A	A
*A. clypeata**	A	A
*A. strepera**	A	A
*A. penelope**	A	A

Species	Breeding habitat	Winter habitat
Aythya ferina	A	A
A. fuligula	A	A
A. nyroca	A	A
ACCIPITRIDAE		
Pandion haliaetus	F	A
Pernis apivorus	F	O
*Milvus migrans***	F	O
*Circaetus gallicus***	F	O
*Circus macrourus**	O	O
*C. pygargus**	O	O
*C. aeruginosus**	O	O
*Neophron percnopterus***	O	O
*Accipiter nisus**	F	F
*Buteo rufinus**	F	O
*B. buteo**	F	O
*Hieraaetus pennatus**	O	O
*Aquila rapax***	O	O
*A. clanga**	O	O
*A. pomarina**	F	O
*A. heliaca**	F	O
*Gyps fulvus***	O	O
FALCONIDAE		
*Falco cherrug**	O	O
*F. vespertinus**	F	O
*F. naumanni**	O	O
*F. subbuteo**	F	F
*F. tinnunculus***	O	O
*F. peregrinus***	O	O
*F. amurensis**	F	F
*F. concolor***	O	O
PHASIANIDAE		
*Coturnix coturnix***	O	O
RALLIDAE		
Crex crex	O	O
*Porzana porzana**	A	A
*P. parva**	A	A
*P. pusilla***	A	A
*Fulica atra**	A	A
GRUIDAE		
Grus grus	O	O
*Anthropoides virgo**	O	A
BURHINIDAE		
*Burhinus oedicnemus**	O	O

Species	Breeding habitat	Winter habitat
GLAREOLIDAE		
*Glareola pratincola***	A	A
*G. nordmanni**	A	A
CHARADRIIDAE		
Vanellus gregarius	O	O
V. leucurus	A	A
Pluvialis squatarola	A	A
Charadrius alexandrinus	A	A
C. hiaticula	A	A
C. dubius	A	A
C. asiaticus	A	A
RECURVIROSTRIDAE		
*Himantopus himantopus***	A	A
*Recurvirostra avosetta***	A	A
SCOLOPACIDAE		
Tringa nebularia	A	A
T. stagnatilis	A	A
T. totanus	A	A
T. erythropus	A	A
T. glareola	A	A
T. ochropus	A	A
T. hypoleucos	A	A
Xenus cinereus	A	A
Numenius phaeopus	A	A
N. arquata	A	A
Limosa limosa	A	A
Arenaria interpres	A	A
Calidris alba	A	A
C. minuta	A	A
C. temminckii	A	A
C. alpina	A	A
C. ferruginea	A	A
Philomachus pugnax	A	A
Lymnocryptes minimus	A	A
*Gallinago gallinago**	A	A
LARIDAE		
*Larus fuscus**	O	A
*Sterna nilotica**	A	A
*S. caspia***	A	A
*Chlidonias hybridus***	A	A
*C. leucopterus**	A	A
COLUMBIDAE		
*Streptopelia turtur***	F	O

Species	Breeding habitat	Winter habitat
CUCULIDAE		
*Cuculus canorus***	F	F
*C. poliocephalus***	F	F
*Clamator glandarius***	F	O
STRIGIDAE		
*Otus scops**	F	F
*Asio flammeus**	O	O
CAPRIMULGIDAE		
*Caprimulgus europaeus**	O	O
*C. ruficollis**	O	O
*C. aegyptius**	O	O
APODIDAE		
*Apus apus***	O	O
*A. melba***	O	O
*A. pallidus***	O	O
*A. affinis***	O	O
UPUPIDAE		
*Upupa epops***	O	O
MEROPIDAE		
*Merops apiaster**	O	O
CORACIIDAE		
*Coracias garrulus**	F	O
PICIDAE		
*Jynx torquilla**	F	O
ALAUDIDAE		
*Calandrella cinerea***	O	O
Melanocorypha bimaculata	O	O
HIRUNDINIDAE		
*Riparia riparia**	O	O
*Hirundo daurica**	O	O
*H. rupestris**	O	O
*H. rustica**	O	O
Delichon urbica	O	O
MUSCICAPIDAE		
Locustella luscinioides	A	A
L. fluviatilis	A	A
L. naevia	O	A
*Acrocephalus melanopogon**	A	A
*A. paludicola**	A	A
*A. schoenobaenus**	A	A

Species	Breeding habitat	Winter habitat
*A. palustris**	A	A
*A. scirpaceus**	A	A
*A. arundinaceus**	A	A
*Hippolais icterina**	F	F
*H. polyglotta**	F	F
*H. olivetorum***	F	O
*H. pallida**	O	O
*H. languida**	F	O
*Sylvia nisoria**	F	O
*S. hortensis**	F	O
*S. borin**	F	O
*S. atricapilla**	F	O
*S. communis**	O	O
*S. curruca**	O	O
*S. rueppelli**	O	O
*S. melanocephala**	F	O
*S. cantillans**	O	O
*S. mystacea**	O	O
*S. nana**	O	O
*Phylloscopus trochilus**	F	F
*P. collybita**	F	F
*P. bonelli**	F	O
*P. sibilatrix**	F	F
Ficedula hypoleuca	F	F
F. albicollis	F	O
*Muscicapa striata***	F	O
*Erithacus luscinia**	O	O
*E. megarhynchos**	O	O
*E. svecica**	O	O
*Oenanthe oenanthe***	O	O
*O. pleschanka**	O	O
*O. hispanica**	O	O
*O. deserti**	O	O
*O. isabellina**	O	O
*O. xanthoprymna**	O	O
*Erythropygia galactotes***	O	O
*Monticola saxatilis**	O	O
*M. solitarius**	O	O
Phoenicurus ochruros	O	O
P. phoenicurus	F	O
*Saxicola rubetra**	O	O
*S. torquata***	O	O
*Turdus philomelos**	F	O
Irania gutturalis	O	O

ORIOLIDAE

*Oriolus oriolus**	F	F

MOTACILLIDAE

*Motacilla flava**	O	O

Species	Breeding habitat	Winter habitat
M. cinerea*	O	O
M. alba*	O	O
Anthus trivialis*	F	O
A. cervinus*	F	O
A. campestris*	O	O
LANIIDAE		
Lanius collurio*	O	O
L. nubicus*	F	O
L. senator*	F	O
L. minor*	F	O
L. excubitor**	O	O
L. isabellinus*	O	O
EMBERIZIDAE		
Emberiza cineracea*	F	O
E. hortulana*	F	O
E. caesia*	F	O
PLOCEIDAE		
Petronia brachydactyla*	O	O

APPENDIX 3: SYSTEMATIC RELATIONSHIPS AND HABITAT USE FOR PALEARCTIC–ASIAN MIGRANTS

The list of Palearctic–Asian migrant species is a synthesis of Peters et al. (1931–1986), Ali and Ripley (1968–1974), Lekagul (1968), Wildash (1968), McClure (1974), and King and Dickinson (1975). Principal habitats are as follows: A = aquatic, F = forest, O = open (i.e., grassland, cropland, desert, savannah, scrub). Breeding habitats are based on information in Vaurie (1959, 1965). Winter habitats are based on information in Ali and Ripley (1968–1974), Lekagul (1968), Wildash (1968), McClure (1974), and King and Dickinson (1975). Those species for which a congener is known to breed in the Asian Tropics are marked with an asterisk (*); those species in which a conspecific is known to breed in the Asian Tropics are marked with two asterisks (**), based on taxonomic data in Peters et al. (1931–1986).

Species	Breeding habitat	Winter habitat
PODICIPEDIDAE		
*Podiceps cristatus**	A	A
PELECANIDAE		
*Pelecanus onocrotalus**	A	A
*P. philippensis***	A	A
PHALACROCORACIDAE		
*Phalacrocorax carbo**	A	A
ARDEIDAE		
Botaurus stellaris	A	A
*Ixobrychus cinnamomeus***	A	A
*I. sinensis***	A	A
*I. eurhythmus**	A	A
*Ardea cinerea***	A	A
*A. purpurea***	A	A
*Casmerodius albus***	A	A
*Egretta eulophotes**	A	A
*E. garzetta***	A	A
*E. intermedia***	A	A
*Bubulcus ibis***	O	O
*Butorides striatus***	A	A
*Ardeola bacchus**	A	A

Species	Breeding habitat	Winter habitat
*Gorsachius goisagi**	F	F
*Nycticorax nycticorax***	A	A
*Dupetor flavicollis***	A	A
THRESKIORNITHIDAE		
*Threskiornis melanocephalus**	A	A
*Plegadis falcinellus***	A	A
*Platalea leucorodia***	A	A
*P. minor***	A	A
CICONIIDAE		
*Ibis leucocephalus***	A	A
*Anastomus oscitans***	A	A
*Ciconia ciconia**	A	A
*C. nigra**	A	A
*C. episcopus***	A	A
*Leptoptilos javanicus***	A	A
ANATIDAE		
Anser anser	A	A
A. indicus	A	A
Tadorna ferruginea	A	A
*Anas crecca**	A	A
*A. acuta**	A	A
*A. poecilorhyncha**	A	A
*A. platyrhynchos**	A	A
*A. querquedula**	A	A
*A. clypeata**	A	A
*A. strepera**	A	A
*A. penelope**	A	A
*A. falcata**	A	A
Rhodonessa caryophyllacea	F	F
Netta rufina	A	A
Aythya ferina	A	A
A. fuligula	A	A
A. nyroca	A	A
A. baeri	A	A
Aix galericulata	A	A
ACCIPITRIDAE		
*Pandion haliaetus***	A	A
*Aviceda leuphotes***	F	F
*Pernis apivorus***	F	F
*Milvus migrans***	O	O
*Circaetus gallicus***	F	F
*Spilornis cheela***	F	F
*Circus macrourus**	O	O
*C. aeruginosus***	O	O
*C. cyaneus**	O	O
*C. pygargus**	O	O

Species	Breeding habitat	Winter habitat
*C. melanoleucos***	O	O
*Accipiter nisus**	O	O
*A. gentilis**	F	F
*A. gularis**	F	F
*A. soloensis**	F	F
*Buteo buteo**	O	O
*Butastur indicus**	F	F
*Aquila rapax***	O	O
*A. clanga**	A	A
*A. heliaca**	O	O
*Hieraaetus pennatus**	F	F
*Spizaetus nipalensis**	F	F
Aegypius monachus	O	O
FALCONIDAE		
*Falco naumanni**	O	O
*F. subbuteo**	O	O
*F. tinnunculus***	O	O
*F. peregrinus***	O	O
*F. columbarius**	O	O
*F. amurensis**	O	O
PHASIANIDAE		
*Coturnix japonica***	O	O
RALLIDAE		
*Rallus aquaticus***	A	A
*Rallina eurizonoides***	F	F
*Porzana fusca***	A	A
*P. pusilla***	A	A
*P. paykullii***	A	A
*Gallicrex cinerea***	A	A
*Gallinula chloropus***	A	A
*Fulica atra**	A	A
GRUIDAE		
Grus grus	A	A
*Anthropoides virgo**	A	A
JACANIDAE		
*Hydrophasianus chirurgus***	A	A
ROSTRATULIDAE		
*Rostratula benghalensis***	A	A
BURHINIDAE		
*Burhinus oedicnemus**	O	O
GLAREOLIDAE		
*Glareola maldivarum***	O	O

Species	Breeding habitat	Winter habitat
CHARADRIIDAE		
*Vanellus vanellus**	O	O
*V. cinereus**	A	A
*V. indicus***	A	A
Pluvialis squatarola	A	A
P. dominica	A	A
*Charadrius dubius***	A	A
*C. alexandrinus**	A	A
*C. placidus**	A	A
*C. mongolus**	A	A
*C. leschenaultii**	A	A
*C. veredus**	A	A
HAEMATOPODIDAE		
Haematopus ostralegus	A	A
RECURVIROSTRIDAE		
*Himantopus himantopus***	A	A
*Recurvirostra avosetta**	A	A
SCOLOPACIDAE		
Tringa nebularia	A	A
T. stagnatilis	A	A
T. totanus	A	A
T. guttifer	A	A
T. erythropus	A	A
T. glareola	A	A
T. ochropus	A	A
T. hypoleucos	A	A
Xenus cinereus	A	A
Numenius phaeopus	A	A
N. arquata	A	A
N. minutus	A	A
N. madagascariensis	A	A
Limosa limosa	A	A
L. lapponica	A	A
Heteroscelus brevipes	A	A
Arenaria interpres	A	A
Limnodromus semipalmatus	A	A
Calidris canutus	A	A
C. tenuirostris	A	A
C. alba	A	A
C. ruficollis	A	A
C. minuta	A	A
C. temminckii	A	A
C. subminuta	A	A
C. acuminata	A	A
C. alpina	A	A
C. ferruginea	A	A
Limicola falcinellus	A	A

Species	Breeding habitat	Winter habitat
Eurynorhynchus pygmaeus	A	A
Philomachus pugnax	A	A
Lymnocryptes minimus	A	A
*Gallinago nemoricola**	A	A
*G. stenura**	A	A
*G. megala**	A	A
*G. gallinago**	A	A
Scolopax rusticola	F	F
Phalaropus lobatus	A	A
LARIDAE		
Larus ichthyaetus	A	A
L. ridibundus	A	A
L. brunnicephalus	A	A
L. argentatus	A	A
*Sterna nilotica**	A	A
*S. caspia**	A	A
*S. aurantia***	A	A
*S. hirundo**	A	A
*S. dougallii***	A	A
*S. sumatrana***	A	A
*S. albifrons***	A	A
*S. saundersi**	A	A
*S. bergii***	A	A
*S. bengalensis**	A	A
*S. zimmermanni**	A	A
*Chlidonias hybridus***	A	A
*C. leucopterus**	A	A
COLUMBIDAE		
*Streptopelia orientalis***	F	F
*S. tranquebarica***	O	O
CUCULIDAE		
*Cuculus canorus***	F	F
*C. sparverioides***	F	F
*C. fugax***	F	F
*C. micropterus***	F	F
*C. saturatus***	F	F
*C. poliocephalus***	F	F
*Clamator coromandus***	F	F
*Chrysococcyx maculatus***	F	F
*Surniculus lugubris***	F	F
STRIGIDAE		
*Ninox scutulata***	F	F
*Asio flammeus**	O	O
CAPRIMULGIDAE		
*Caprimulgus indicus***	O	F

Species	Breeding habitat	Winter habitat
APODIDAE		
*Collocalia brevirostris***	O	O
*Hirundapus caudacutus**	O	O
*Apus pacificus***	F	F
*A. affinis***	O	O
UPUPIDAE		
*Upupa epops***	O	O
ALCEDINIDAE		
*Ceyx erithacus***	F	F
*Halcyon coromanda***	F	F
MEROPIDAE		
*Merops philippinus***	O	O
*M. viridis***	O	O
CORACIIDAE		
*Eurystomus orientalis***	F	F
PICIDAE		
*Jynx torquilla**	O	O
ALAUDIDAE		
*Mirafra javanica***	O	O
*Calandrella cinerea**	O	O
HIRUNDINIDAE		
*Riparia riparia**	A	A
*Hirundo daurica***	O	O
*H. rustica**	O	O
Delichon urbica	F	F
D. dasypus	F	F
CAMPEPHAGIDAE		
*Coracina melaschista***	F	F
*C. melanoptera**	F	F
*Pericrocotus divaricatus**	F	F
*P. roseus**	F	F
*P. ethologus***	F	F
PYCNONOTIDAE		
*Hypsipetes madagascariensis***	F	F
DICRURIDAE		
*Dicrurus macrocercus***	O	O
*D. leucophaeus***	F	F
*D. annectans***	F	F
*D. hottentottus***	F	F

Species	Breeding habitat	Winter habitat
ORIOLIDAE		
*Oriolus chinensis***	O	O
*O. tenuirostris***	O	O
*O. xanthornus***	F	F
*O. mellianus**	F	F
MUSCICAPIDAE		
*Cettia squameiceps**	F	F
*C. pallidipes***	O	O
*C. canturians**	O	O
*Bradypterus thoracicus**	O	O
*B. tacsanowskius**	O	O
*B. luteoventris**	O	O
Locustella lanceolata	A	A
L. certhiola	A	A
L. ochotensis	A	A
L. fasciolata	A?	A
L. amnicola	A?	A
*Acrocephalus arundinaceus**	A	A
*A. bistrigiceps**	A	A
*A. agricola**	A	A
*A. concinens**	A	A
*A. dumetorum**	A	A
*A. stentoreus***	A	A
*A. aedon**	O	O
*Sylvia hortensis**	O	O
*S. curruca**	O	O
*Phylloscopus affinis**	O	O
*P. subaffinis**	O	O
*P. fuscatus**	A	O
*P. armandii**	O	O
*P. schwarzi**	O	O
*P. pulcher**	F	F
*P. maculipennis**	F	F
*P. proregulus**	F	F
*P. inornatus**	F	F
*P. borealis**	F	F
*P. trochiloides**	F	F
*P. plumbeitarsus**	F?	F
*P. tenellipes**	F	F
*P. magnirostris**	F	F
*P. occipitalis**	F	F
*P. coronatus**	F	F
*P. ijimae**	F?	F
*P. reguloides**	F	F
*P. davisoni**	F	F
*P. cantator**	F	F
*P. ricketti**	F	F
*Seicercus burkii**	F	F

Species	Breeding habitat	Winter habitat
*Abroscopus superciliaris***	F	F
*A. albogularis***	F	F
*Rhinomyias brunneata**	F	F
*Muscicapa griseisticta**	F	F
*M. sibirica**	F	F
*M. latirostris**	F	F
*M. ferruginea**	F	F
*M. thalassina***	F	F
*Ficedula zanthopygia**	F	F
*F. narcissina**	F	F
*F. mugimaki**	F	F
*F. parva**	O	O
*F. strophiata**	F	F
*F. hodgsonii**	F	F
*F. superciliaris**	F	F
Cyanoptila cyanomelana	F	F
*Niltava sundara***	F	F
*Cyornis rubeculoides***	F	F
*C. hainana***	F	F
*Culicicapa ceylonensis***	F	F
*Hypothymis azurea***	O	O
*Terpsiphone atrocaudata**	F	F
*T. paradisi***	F	F
*Erithacus akahige**	F	F
*E. sibilans**	F	F
*E. calliope**	F	F
*E. svecica**	O	O
*E. cyane**	F	F
*Tarsiger cyanurus**	F	F
*T. chrysaeus***	O	O
*Monticola cinclorhynchus**	F	F
*M. gularis**	F	F
*M. rufiventris***	F	F
*M. solitarius***	O	O
*Myophonus caeruleus***	F	F
*Zoothera citrina***	F	F
*Z. sibirica**	F	F
*Z. mollissima**	F	F
*Z. dauma***	F	F
Phoenicurus ochruros	O	O
P. frontalis	O	O
P. auroreus	O	O
*Rhyacornis fuliginosus***	A	A
*Saxicola torquata***	O	O
*S. ferrea**	O	O
*Turdus hortulorum**	F	F
*T. cardis**	F	F
*T. merula**	O	O
*T. chrysolaus**	F	F
*T. rubrocanus**	F	F

Species	Breeding habitat	Winter habitat
*T. feae**	F	F
*T. pallidus**	F	F
*T. obscurus**	F	F
*T. ruficollis**	F	O
*T. naumanni**	O	O
MOTACILLIDAE		
*Motacilla flava**	O	O
*M. cinerea**	O	O
*M. alba**	O	O
*M. citreola**	O	O
Dendronanthus indicus	A	F
*Anthus hodgsoni**	F	F
*A. cervinus**	O	O
*A. novaeseelandiae***	O	O
*A. godlewskii**	O	O
*A. gustavi**	O	O
*A. roseatus**	O	O
*A. similis**	O	O
LANIIDAE		
*Lanius cristatus**	O	O
*L. tigrinus**	F	F
*L. tephronotus**	O	O
STURNIDAE		
Saroglossa spiloptera	O	O
*Sturnus malabaricus***	O	O
*S. sericeus***	O	O
*S. sinensis**	O	O
*S. sturninus**	O	O
*S. cineraceus**	O	O
ZOSTEROPIDAE		
*Zosterops erythropleura**	O	O
*Z. palpebrosa**	O	F
EMBERIZIDAE		
Emberiza tristrami	F	F
E. fucata	O	O
E. pusilla	O	O
E. aureola	O	O
E. rutila	F	F
E. spodocephala	O	O
FRINGILLIDAE		
Carduelis ambigua	O	O
Carpodacus nipalensis	O	F
C. erythrinus	O	O
*Coccothraustes migratorius**	F	F

APPENDIX 4: ENGLISH COMMON NAMES AND THEIR COUNTERPARTS

English common name	Scientific name	Spanish common name
Anhinga	*Anhinga anhinga*	Anhinga Americana
Ani, Groove-billed	*Crotophaga sulcirostris*	—
Antthrush, Black-faced	*Formicarius analis*	—
Avocet, American	*Recurvirostra americana*	Avoceta Piquicurva
Bittern		
American	*Botaurus lentiginosus*	Garza Norteña de Tular
Least	*Ixobrychus exilis*	Garcita de Tular
Blackbird		
Brewer's	*Euphagus cyanocephalus*	Tordo Ojiclaro
Melodious	*Dives dives*	—
Red-winged	*Agelaius phoeniceus*	Tordo Sargento
Yellow-headed	*Xanthocephalus xanthocephalus*	Tordo Cabeciamarillo
Black-Hawk, Common	*Buteogallus anthracinus*	Aguililla Negra Menor
Bluebird		
Eastern	*Sialia sialis*	Azulejo Gorjicanelo
Mountain	*S. currucoides*	Azulejo Pálido
Western	*S. mexicana*	Azulejo Gorjiazul
Bobolink	*Dolichonyx oryzivorus*	Tordo Migratorio
Bunting		
Indigo	*Passerina cyanea*	Colorín Azul
Lark	*Calamospiza melanocorys*	Llanero Alipálido
Lazuli	*Passerina amoena*	Colorín Aliblanco
Painted	*P. ciris*	Colorín Sietecolores
Varied	*P. versicolor*	Colorín Oscuro
Canvasback	*Aythya valisineria*	Pato Borrado
Cardinal, Northern	*Cardinalis cardinalis*	
Catbird, Gray	*Dumetella carolinensis*	Mímido Gris
Chat, Yellow-breasted	*Icteria virens*	Chipe Piquigrueso
Chlorophonia, Blue-crowned	*Chlorophonia occipitalis*	—
Chuck-will's-widow	*Caprimulgus carolinensis*	Tapacamino de Paso
Coot, American	*Fulica americana*	Gallareta Americana
Cormorant		
Double-crested	*Phalacrocorax auritus*	Cormorán Orejudo
Olivaceous	*P. olivaceus*	Cormorán Oliváceo

English common name	Scientific name	Spanish common name
Cowbird		
Bronzed	*Molothrus aeneus*	Tordo Ojirrojo
Brown-headed	*M. ater*	Tordo Cabecicafé
Crane		
Sandhill	*Grus canadensis*	Grulla Gris
Whooping	*G. americana*	Grulla Blanca
Cuckoo		
Black-billed	*Coccyzus erythropthalmus*	Cuclillo Piquinegro
Mangrove	*C. minor*	Cuclillo Ventrisucio
Pheasant	*Dromococcyx phasianellus*	—
Yellow-billed	*Coccyzus americanus*	Cuclillo Alirrojizo
Curassow, Great	*Crax rubra*	—
Curlew		
Eskimo	*Numenius borealis*	Zarapito Boreal
Long-billed	*N. americanus*	Zarapito Piquilargo
Dickcissel	*Spiza americana*	Espiza
Dove		
Inca	*Columbina inca*	—
Mourning	*Zenaida macroura*	Paloma Común
White-winged	*Z. asiatica*	Paloma Aliblanca
Dowitcher		
Long-billed	*Limnodromus scolopaceus*	Costurero de Agua Dulce
Short-billed	*L. griseus*	Costurero Marino
Duck		
Ring-necked	*Aythya collaris*	Pato Piquianillado
Ruddy	*Oxyura jamaicensis*	Pato Rojizo Alioscuro
Wood	*Aix sponsa*	Pato Arcoiris
Egret		
Cattle	*Bubulcus ibis*	Garza Ganadera
Great	*Casmerodius albus*	Garzón Blanco
Reddish	*Egretta rufescens*	Garza Piquirrosa
Snowy	*E. thula*	Garza Dedos Dorados
Euphonia, Yellow-throated	*Euphonia hirundinacea*	—
Falcon		
Peregrine	*Falco peregrinus*	Halcón Peregrino
Prairie	*F. mexicanus*	Halcón Pálido
Sooty	*F. concolor*	—
Flycatcher		
Acadian	*Empidonax virescens*	Empidonax Verdoso
Alder	*E. alnorum*	Empidonax Alnoro
Ash-throated	*Myiarchus cinerascens*	Papamoscas Copetón Gorjicenizo
Brown-crested	*M. tyrannulus*	Papamoscas Copetón Tiranillo
Buff-breasted	*Empidonax fulvifrons*	Empidonax Canelo
Collared	*Ficedula albicollis*	
Cordilleran	*Empidonax occidentalis*	Empidonax Occidental
Dusky	*E. oberholseri*	Empidonax de Oberholser
Dusky-capped	*Myiarchus tuberculifer*	Papamoscas Copetón Triste

English common name	Scientific name	Spanish common name
Gray	*Empidonax wrightii*	Empidonax de Wright
Great Crested	*Myiarchus crinitus*	Papamoscas Copetón Viajero
Hammond's	*Empidonax hammondii*	Empidonax de Hammond
Least	*E. minimus*	Empidonax Mínimo
Olive-sided	*Contopus borealis*	Contopus de Chaleco
Pacific-slope	*Empidonax difficilis*	Empidonax Difícil
Pied	*Ficedula hypoleuca*	—
Royal	*Onychorhynchus coronatus*	—
Scissor-tailed	*Tyrannus forficatus*	Tirano Tijereta Clara
Social	*Myiozetetes similis*	—
Sulphur-bellied	*Myiodynastes luteiventris*	Papamoscas Rayado Cejiblanco
Sulphur-rumped	*Myiobius sulphureipygius*	—
Vermilion	*Pyrocephalus rubinus*	Mosquero Cardenalito
Willow	*Empidonax traillii*	Empidonax de Traill
Yellow-bellied	*E. flaviventris*	Empidonax Ventriamarillo
Gadwall	*Anas strepera*	Pato Friso
Gallinule, Purple	*Porphyrula martinica*	Gallareta Morada
Gnatcatcher, Blue-gray	*Polioptila caerulea*	Perlita Piis
Godwit		
Hudsonian	*Limosa haemastica*	Limosa Ornamentada
Marbled	*L. fedoa*	Limosa Canela
Goldfinch		
American	*Carduelis tristis*	Jilguero Canario
Lesser	*C. psaltria*	Jilguero Dorsioscuro
Goose		
Greater White-fronted	*Anser albifrons*	Ganso Manchado
Snow	*Chen caerulescens*	Ganso Cerúleo (Blanco)
Grebe		
Clark's	*Aechmophorus clarkii*	Zambullidor de Clark
Eared	*Podiceps nigricollis*	Zambullidor Mediano
Least	*Tachybaptus dominicus*	—
Pied-billed	*Podilymbus podiceps*	Zambullidor Piquigrueso
Western	*Aechmophorus occidentalis*	Zambullidor Achichilique
Greenlet, Lesser	*Hylophilus decurtatus*	—
Grosbeak		
Black-headed	*Pheucticus melanocephalus*	Picogrueso Pechicafé
Blue	*Guiraca caerulea*	Picogrueso Azul
Rose-breasted	*Pheucticus ludovicianus*	Picogrueso Pechirrosa
Gull		
Bonaparte's	*Larus philadelphia*	Gaviota Menor
California	*L. californicus*	Gaviota Californiana
Franklin's	*L. pipixcan*	Gaviota Apipizca
Glaucous-winged	*L. glaucescens*	Gaviota Aliglauca
Herring	*L. argentatus*	Gaviota Plateada
Laughing	*L. atricilla*	Gaviota Atricila
Ring-billed	*L. delawarensis*	Gaviota de Delaware
Sabine's	*Xema sabini*	Gaviota Colihendida
Western	*Larus occidentalis*	Gaviota Occidental
Harrier, Northern	*Circus cyaneus*	Aguililla Rastrera

English common name	Scientific name	Spanish common name
Hawk		
Broad-winged	*Buteo platypterus*	Aguililla Migratoria Menor
Cooper's	*Accipiter cooperii*	Gavilán Pechirrufo Mayor
Ferruginous	*Buteo regalis*	Aguililla Real
Great Black-	*Buteogallus urubitinga*	—
Red-tailed	*Buteo jamaicensis*	Aguililla Colirrufa
Sharp-shinned	*Accipiter striatus*	Gavilán Pechirrufo Menor
Short-tailed	*Buteo brachyurus*	—
Swainson's	*B. swainsoni*	Aguililla Migratoria Mayor
White-tailed	*B. albicaudatus*	—
Heron		
Great Blue	*Ardea herodias*	Garzón Cenizo
Green-backed	*Butorides striatus*	Garcita Oscura
Little Blue	*Egretta caerulea*	Garza Azul
Tricolored	*E. tricolor*	Garza Ventriblanca
Hummingbird		
Allen's	*Selasphorus sasin*	Colibrí Colicanelo Sasin
Anna's	*Calypte anna*	Colibrí Coronirrojo
Black-chinned	*Archilochus alexandri*	Colibrí Gorjinegro
Blue-throated	*Lampornis clemenciae*	Chupaflor Gorjiazul
Broad-billed	*Cynanthus latirostris*	Colibrí Latirrostro
Broad-tailed	*Selasphorus platycercus*	Colibrí Vibrador
Buff-bellied	*Amazilia yucatanensis*	Amazilia del Golfo
Calliope	*Stellula calliope*	Colibrí Gorjirrayado
Costa's	*Calypte costae*	Colibrí Coronivioleta
Magnificent	*Eugenes fulgens*	Chupaflor Coronimorado
Ruby-throated	*Archilochus colubris*	Colibrí de Paso
Rufous	*Selasphorus rufus*	Colibrí Colicanelo Rufo
Violet-crowned	*Amazilia violiceps*	Amazilia Occidental
Ibis		
Glossy	*Plegadis falcinellus*	Ibis Negro
White	*Eudocimus albus*	Ibis Blanco
White-faced	*Plegadis chihi*	Ibis Cara Blanca
Kestrel, American	*Falco sparverius*	Halcón Cernícalo
Killdeer	*Charadrius vociferus*	Chorlito Tildío
Kingbird		
Cassin's	*Tyrannus vociferans*	Tirano Gritón
Couch's	*T. couchii*	Tirano Tropical Silbador
Eastern	*T. tyrannus*	Tirano Dorsinegro
Gray	*T. dominicensis*	Tirano Dominicano
Thick-billed	*T. crassirostris*	Tirano Piquigrueso
Tropical	*T. melancholicus*	Tirano Tropical Común
Western	*T. verticalis*	Tirano Pálido
Kingfisher, Belted	*Ceryle alcyon*	Martín Pescador Norteño
Kinglet, Ruby-crowned	*Regulus calendula*	Reyezuelo Sencillo
Kite		
American Swallow-tailed	*Elanoides forficatus*	Milano Tijereta
Mississippi	*Ictinia mississippiensis*	Milano Migratorio
Snail	*Rostrhamus sociabilis*	—

English common name	Scientific name	Spanish common name
Knot, Red	*Calidris canutus*	Playero Piquicorto
Mallard	*Anas platyrhynchos*	Pato de Collar
Martin, Purple	*Progne subis*	Golondrina Grande Negruzca
Meadowlark		
Eastern	*Sturnella magna*	Pradero Tortilla-con-chile
Western	*S. neglecta*	Pradero Gorjeador
Merganser		
Hooded	*Lophodytes cucullatus*	Pato Mergo Copetón
Red-breasted	*Mergus serrator*	Pato Mergo Pechicastaño
Merlin	*Falco columbarius*	Halcón Esmerejón
Mockingbird		
Northern	*Mimus polyglottos*	—
Tropical	*M. gilvus*	—
Moorhen, Common	*Gallinula chloropus*	Gallareta Frentirroja
Nighthawk		
Common	*Chordeiles minor*	Chotacabra Zumbón
Lesser	*C. acutipennis*	Chotacabra Halcón
Night-Heron		
Black-crowned	*Nycticorax nycticorax*	Garza Nocturna Coroninegra
Yellow-crowned	*Nyctanassa violacea*	Garza Nocturna Coroniclara
Oriole		
Audubon's	*Icterus graduacauda*	Bolsero Capuchinegro
"Baltimore"	*I. galbula*	—
"Bullock's"	*I. galbula*	—
Hooded	*I. cucullatus*	Bolsero Cuculado
Northern	*I. galbula*	Bolsero Norteño
Orchard	*I. spurius*	Bolsero Castaño
Scott's	*I. parisorum*	Bolsero Parisino
Osprey	*Pandion haliaetus*	Aguila Pescadora
Ovenbird	*Seiurus aurocapillus*	Chipe Suelero Coronado
Owl		
Burrowing	*Speotyto cunicularia*	Lechucilla Zancón
Elf	*Micrathene whitneyi*	Lechucilla Colicorta
Short-eared	*Asio flammeus*	Buho Cornicorto Llanero
Oystercatcher, American	*Haematopus palliatus*	Ostrero Blanquinegro
Parula, Northern	*Parula americana*	Chipe Azul-Olivo Norteño
Pauraque, Common	*Nyctidromus albicollis*	—
Pelican		
American White	*Pelecanus erythrorhynchos*	Pelícano Blanco
Brown	*P. occidentalis*	Pelícano Pardo
Pewee		
Eastern Wood-	*Contopus virens*	Contopus Verdoso
Greater	*C. pertinax*	Contopus José María
Western Wood-	*C. sordidulus*	Contopus Occidental
Phalarope		
Red	*Phalaropus fulicaria*	Falaropo Piquigrueso
Red-necked	*P. lobatus*	Falaropo Piquifino
Wilson's	*P. tricolor*	Falaropo Piquilargo
Phoebe		
Eastern	*Sayornis phoebe*	Mosquero Fibí
Say's	*S. saya*	Mosquero Llanero

English common name	Scientific name	Spanish common name
Pigeon		
Band-tailed	*Columba fasciata*	Paloma Collareja
Red-billed	*C. flavirostris*	Paloma Morada Ventrioscura
White-crowned	*C. leucocephala*	Paloma Cabeciblanca
Pintail, Northern	*Anas acuta*	Pato Golondrino
Pipit		
American	*Anthus rubescens*	Bisbita Americano
Sprague's	*A. spragueii*	Bisbita Llanera
Pitta, Hooded	*Pitta sordida*	—
Plover		
Black-bellied	*Pluvialis squatarola*	Chorlo Axilinegro
Lesser Golden-	*P. dominica*	Chorlo Axiliclaro
Mountain	*Charadrius montanus*	Chorlito Llanero
Piping	*C. melodus*	Chorlito Melódico
Semipalmated	*C. semipalmatus*	Playerito Semipalmeado
Snowy	*C. alexandrinus*	Chorlito Alejandrino
Wilson's	*C. wilsonia*	Chorlito Piquigrueso
Poorwill, Common	*Phalaenoptilus nuttallii*	Tapacamino Tevíi
Quetzal, Resplendent	*Pharomachrus mocinno*	—
Rail		
Black	*Laterallus jamaicensis*	Ralito Negruzco
King	*Rallus elegans*	Ralón Barrado Rojizo
Virginia	*R. limicola*	Ralo Barrado Rojizo
Redhead	*Aythya americana*	Pato Cabecirrojo
Redstart		
American	*Setophaga ruticilla*	Pavito Migratorio
Painted	*Myioborus pictus*	Pavito Aliblanco
Slate-throated	*M. miniatus*	
Robin		
American	*Turdus migratorius*	Zorzal Pechirrojo
European	*Erithacus rubecula*	—
White-throated	*Turdus assimilis*	—
Sanderling	*Calidris alba*	Playerito Correlón
Sandpiper		
Baird's	*Calidris bairdii*	Playerito de Baird
Buff-breasted	*Tryngites subruficollis*	Playerito Pradero
Least	*Calidris minutilla*	Playerito Mínimo
Pectoral	*C. melanotos*	Playero Pechirrayado
Semipalmated	*C. pusilla*	Playerito Semipalmeado
Solitary	*Tringa solitaria*	Playero Charquero
Spotted	*Actitis macularia*	Playerito Alzacolita
Stilt	*Calidris himantopus*	Playero Zancón
Upland	*Bartramia longicauda*	Zarapito Ganga
Western	*Calidris mauri*	Playerito Occidental
White-rumped	*C. fuscicollis*	Playerito de Rabadilla Blanca
Sapsucker		
Red-breasted	*Sphyrapicus thyroideus*	Carpintero Aliblanco Oscuro
Red-naped	*S. nuchalis*	Carpintero Nuquirrojo
Yellow-bellied	*S. varius*	Carpintero Aliblanco Común
Scaup, Lesser	*Aythya affinis*	Pato Boludo Menor

English common name	Scientific name	Spanish common name
Shoveler, Northern	*Anas clypeata*	Pato Cucharón
Shrike, Loggerhead	*Lanius ludovicianus*	Verdugo Americano
Skimmer, Black	*Rynchops niger*	Rayador Americano
Snipe, Common	*Gallinago gallinago*	Agachona Común
Solitaire		
Slate-colored	*Myadestes unicolor*	—
Townsend's	*M. townsendi*	Clarín Norteño
Sora	*Porzana carolina*	Ralo Barrado Grisáceo
Sparrow		
Bachman's	*Aimophila aestivalis*	—
Black-chinned	*Spizella atrogularis*	Gorrión Indefinido Oriental
Botteri's	*Aimophila botterii*	Gorrión de Botteri
Brewer's	*Spizella breweri*	Gorrión de Brewer
Chipping	*S. passerina*	Gorrión Llanero
Clay-colored	*S. pallida*	Gorrión Indefinido Rayado
Grasshopper	*Ammodramus savannarum*	Gorrión Saltamonte
Lark	*Chondestes grammacus*	Gorrión Arlequín
Lincoln's	*Melospiza lincolnii*	Gorrión de Lincoln
Rufous-crowned	*Aimophila ruficeps*	Gorrión Bigotudo Coronirrufo
Rufous-winged	*A. carpalis*	Gorrión Bigotudo Sonorense
Savannah	*Passerculus sandwichensis*	Gorrión Sabanero Común
Song	*Melospiza melodia*	—
Swamp	*M. georgiana*	Gorrión Georgiana
Vesper	*Pooecetes gramineus*	Gorrión Zacatero Coliblanco
White-crowned	*Zonotrichia leucophrys*	Gorrión Gorriblanco
Spoonbill, Roseate	*Ajaia ajaja*	Ibis Espátula
Stilt, Black-necked	*Himantopus mexicanus*	Avoceta Piquirrecta
Stork, Wood	*Mycteria americana*	Cigüeña Americana
Surfbird	*Aphriza virgata*	Playero Roquero
Swallow		
Bank	*Riparia riparia*	Golondrina Pechifajada
Barn	*Hirundo rustica*	Golondrina Tijereta
Cave	*H. fulva*	Golondrina Fulva
Cliff	*H. pyrrhonota*	Golondrina Risquera
Northern Rough-winged	*Stelgidopteryx serripennis*	Golondrina Gorjicafé
Tree	*Tachycineta bicolor*	Golondrina Canadiense
Violet-green	*T. thalassina*	Golondrina Cariblanca
Swift		
Black	*Cypseloides niger*	Vencejo Negro
Chimney	*Chaetura pelagica*	Vencejito de Paso
Vaux's	*C. vauxi*	Vencejito Alirrápido
White-throated	*Aeronautes saxatalis*	Vencejito Pechiblanco
Tanager		
Common Bush-	*Chlorospingus ophthalmicus*	—
Hepatic	*Piranga flava*	Tangara Roja Piquioscura
Scarlet	*P. olivacea*	Tangara Rojinegra Migratoria
Summer	*P. rubra*	Tangara Roja Migratoria
Western	*P. ludoviciana*	Tangara Aliblanca Migratoria
Tattler, Wandering	*Heteroscelus incanus*	Playero Sencillo

English common name	Scientific name	Spanish common name
Teal		
Blue-winged	*Anas discors*	Cerceta Aliazul Clara
Cinnamon	*A. cyanoptera*	Cerceta Aliazul Café
Green-winged	*A. crecca*	Cerceta Alioscura
Tern		
Black	*Chlidonias niger*	Gaviotina Negruzca
Caspian	*Sterna caspia*	Gaviotina Piquirroja
Common	*S. hirundo*	Gaviotina Común
Elegant	*S. elegans*	Gaviotina Elegante
Forster's	*S. forsteri*	Gaviotina de Forster
Gull-billed	*S. nilotica*	Gaviotina Piquigruesa
Least	*S. antillarum*	Gaviotina Menor
Roseate	*S. dougallii*	Gaviotina de Dougall
Royal	*S. maxima*	Gaviotina Piquinaranja
Sandwich	*S. sandvicensis*	Gaviotina de Sandwich
Sooty	*S. fuscata*	Gaviotina Dorsinegra
Thrasher		
Brown	*Toxostoma rufum*	—
Sage	*Oreoscoptes montanus*	Mímido Pinto
Thrush		
Gray-cheeked	*Catharus minimus*	Zorzalito Carigrís
Hermit	*C. guttatus*	Zorzalito Colirrufo
Swainson's	*C. ustulatus*	Zorzalito de Swainson
Wood	*C. mustelinus*	Zorzalito Maculado
Towhee		
Green-tailed	*Pipilo chlorurus*	Rascador Migratorio
Rufous-sided	*P. erythrophthalmus*	Rascador Pinto Oscuro
Trogon, Elegant	*Trogon elegans*	Trogón Colicobrizo
Turnstone, Ruddy	*Arenaria interpres*	Vuelvepiedras Común
Tyrannulet, Northern Beardless	*Camptostoma imberbe*	Mosquerito Silbador
Veery	*Catharus fuscescens*	Zorzalito de Wilson
Vireo		
Bell's	*Vireo bellii*	Vireo de Bell
Black-capped	*V. atricapillus*	Vireo Gorrinegro
Black-whiskered	*V. altiloquus*	Vireo Bigotinegro
Gray	*V. vicinior*	Vireo Gris
Least Bell's	*V. bellii*	—
Philadelphia	*V. philadelphicus*	Vireo Filadélfico
Red-eyed	*V. olivaceus*	Vireo Ojirrojo Norteño
Solitary	*V. solitarius*	Vireo Anteojillo
Warbling	*V. gilvus*	Vireo Gorjeador Norteño
White-eyed	*V. griseus*	Vireo Ojiblanco
Yellow-green	*V. flavoviridis*	Vireo Ojirrojo Tropical
Yellow-throated	*V. flavifrons*	Vireo Pechiamarillo
Vulture		
Black	*Coragyps atratus*	Carroñero Común
Lesser Yellow-headed	*Cathartes burrovianus*	—
Turkey	*C. aura*	Aura Común

English common name	Scientific name	Spanish common name
Warbler		
Arctic	*Phylloscopus borealis*	—
Bachman's	*Vermivora bachmanii*	Chipe de Bachman
Bay-breasted	*Dendroica castanea*	Chipe Pechicastaño
Blackburnian	*D. fusca*	Chipe Gorjinaranja
Blackpoll	*D. striata*	Chipe Gorrinegro
Black-and-white	*Mniotilta varia*	Chipe Trepador
Black-throated Blue	*Dendroica caerulescens*	Chipe Azul Pizarra
Black-throated Gray	*D. nigrescens*	Chipe Negrigrís
Black-throated Green	*D. virens*	Chipe Dorsiverde
Blue-winged	*Vermivora pinus*	Chipe Aliazul
Canada	*Wilsonia canadensis*	Chipe de Collar
Cape May	*Dendroica tigrina*	Chipe Tigrino
Cerulean	*D. cerulea*	Chipe Cerúleo
Chestnut-sided	*D. pensylvanica*	Chipe Gorriamarillo
Colima	*Vermivora crissalis*	Chipe de Colima
Connecticut	*Oporornis agilis*	Chipe Cabecigrís Ojianillado
Golden-cheeked	*Dendroica chrysoparia*	Chipe Dorsinegro
Golden-winged	*Vermivora chrysoptera*	Chipe Alidorado
Grace's	*Dendroica graciae*	Chipe Pinero Gorjiamarillo
Great Reed	*Acrocephalus arundinaceus*	—
Hermit	*Dendroica occidentalis*	Chipe Occidental
Hooded	*Wilsonia citrina*	Chipe Encapuchado
Kentucky	*Oporornis formosus*	Chipe Cachetinegro
Kirtland's	*Dendroica kirtlandii*	Chipe de Kirtland
Lucy's	*Vermivora luciae*	Chipe de Lucy
MacGillivray's	*Oporornis tolmiei*	Chipe Cabecigrís de Tolmie
Magnolia	*Dendroica magnolia*	Chipe Colifajado
Mourning	*Oporornis philadelphia*	Chipe Cabecigrís Filadélfico
Nashville	*Vermivora ruficapilla*	Chipe de Nashville
Orange-crowned	*V. celata*	Chipe Celato
Oriental Great Reed	*Acrocephalus orientalis*	—
Palm	*Dendroica palmarum*	Chipe Playero
Pine	*D. pinus*	Chipe Nororiental
Prairie	*D. discolor*	Chipe Galán
Prothonotary	*Protonotaria citrea*	Chipe Cabecidorado
Red-faced	*Cardellina rubrifrons*	Chipe Carirrojo
Swainson's	*Limnothlypis swainsonii*	Chipe Coronicafé
Tennessee	*Vermivora peregrina*	Chipe Peregrino
Townsend's	*Dendroica townsendi*	Chipe de Townsend
Virginia's	*Vermivora virginiae*	Chipe de Virginia
Wilson's	*Wilsonia pusilla*	Chipe Coroninegro
Wood	*Phylloscopus sibilatrix*	—
Worm-eating	*Helmitheros vermivorus*	Chipe Vermívoro
Yellow	*Dendroica petechia*	Chipe Amarillo Norteño
Yellow-rumped	*D. coronata*	Chipe Grupidorado
Yellow-throated	*D. dominica*	Chipe Dominico
Waterthrush		
Louisiana	*Seiurus motacilla*	Chipe Suelero Gorjiblanco

English common name	Scientific name	Spanish common name
Northern	*S. noveboracensis*	Chipe Suelero Gorjijaspeado
Waxwing, Cedar	*Bombycilla cedrorum*	Ampelis Americano
Wheatear, Northern	*Oenanthe oenanthe*	—
Whimbrel	*Numenius phaeopus*	Zarapito Cabecirrayado
Whip-poor-will	*Caprimulgus vociferus*	Tapacamino Cuerporruín
Whistling-Duck		
Black-bellied	*Dendrocygna autumnalis*	Pato Pijije Aliblanco
Fulvous	*D. bicolor*	Pato Pijije Alioscuro
Wigeon, American	*Anas americana*	Pato Chalcuán
Willet	*Catoptrophorus semipalmatus*	Playero Pihuihui
Woodcreeper, Tawny-winged	*Dendrocincla anabatina*	—
Woodpecker, Red-headed	*Melanerpes erythrocephalus*	—
Wren		
Bewick's	*Thryomanes bewickii*	—
Carolina	*Thryothorus ludovicianus*	—
House	*Troglodytes aedon*	Troglodita Continental
Marsh	*Cistothorus palustris*	Troglodita Pantanera
Sedge	*C. platensis*	Troglodita de Ciénega
Yellowlegs		
Greater	*Tringa melanoleuca*	Patamarilla Mayor
Lesser	*T. flavipes*	Patamarilla Menor
Yellowthroat, Common	*Geothlypis trichas*	Mascarita Norteña

APPENDIX 5: SPANISH COMMON NAMES AND THEIR COUNTERPARTS

Spanish common name	Scientific name	English common name
Agachona Común	*Gallinago gallinago*	Common Snipe
Aguila Pescadora	*Pandion haliaetus*	Osprey
Aguililla		
Colirrufa	*Buteo jamaicensis*	Red-tailed Hawk
Migratoria Mayor	*B. swainsoni*	Swainson's Hawk
Migratoria Menor	*B. platypterus*	Broad-winged Hawk
Negra Menor	*Buteogallus anthracinus*	Common Black-Hawk
Rastrera	*Circus cyaneus*	Northern Harrier
Real	*Buteo regalis*	Ferruginous Hawk
Amazilia		
del Golfo	*Amazilia yucatanensis*	Buff-bellied Hummingbird
Occidental	*A. violiceps*	Violet-crowned Hummingbird
Ampelis Americano	*Bombycilla cedrorum*	Cedar Waxwing
Anhinga Americana	*Anhinga anhinga*	Anhinga
Aura Común	*Cathartes aura*	Turkey Vulture
Avoceta		
Piquicurva	*Recurvirostra americana*	American Avocet
Piquirrecta	*Himantopus mexicanus*	Black-necked Stilt
Azulejo		
Gorjicanelo	*Sialia sialis*	Eastern Bluebird
Gorjiazul	*S. mexicana*	Western Bluebird
Pálido	*S. currucoides*	Mountain Bluebird
Bisbita		
Americano	*Anthus rubescens*	American Pipit
Llanera	*A. spragueii*	Sprague's Pipit
Bolsero		
Capuchinegro	*Icterus graduacauda*	Audubon's Oriole
Castaño	*I. spurius*	Orchard Oriole
Cuculado	*I. cucullatus*	Hooded Oriole
Norteño	*I. galbula*	Northern Oriole
Parisino	*I. parisorum*	Scott's Oriole
Buho Cornicorto Llanero	*Asio flammeus*	Short-eared Owl
Carpintero		
Aliblanco Común	*Sphyrapicus varius*	Yellow-bellied Sapsucker
Aliblanco Oscuro	*S. thyroideus*	Red-breasted Sapsucker
Nuquirrojo	*S. nuchalis*	Red-naped Sapsucker
Carroñero Común	*Coragyps atratus*	Black Vulture

Spanish common name	Scientific name	English common name
Cerceta		
Aliazul Café	*Anas cyanoptera*	Cinnamon Teal
Aliazul Clara	*A. discors*	Blue-winged Teal
Alioscura	*A. crecca*	Green-winged Teal
Cigüeña Americana	*Mycteria americana*	Wood Stork
Clarín Norteño	*Myadestes townsendi*	Townsend's Solitaire
Colibrí		
Colicanelo Rufo	*Selasphorus rufus*	Rufous Hummingbird
Colicanelo Sasin	*S. sasin*	Allen's Hummingbird
Coronirrojo	*Calypte anna*	Anna's Hummingbird
Coronivioleta	*C. costae*	Costa's Hummingbird
Gorjinegro	*Archilochus alexandri*	Black-chinned Hummingbird
Gorjirrayado	*Stellula calliope*	Calliope Hummingbird
Latirrostro	*Cynanthus latirostris*	Broad-billed Hummingbird
de Paso	*Archilochus colubris*	Ruby-throated Hummingbird
Vibrador	*Selasphorus platycercus*	Broad-tailed Hummingbird
Colorín		
Aliblanco	*Passerina amoena*	Lazuli Bunting
Azul	*P. cyanea*	Indigo Bunting
Oscuro	*P. versicolor*	Varied Bunting
Sietecolores	*P. ciris*	Painted Bunting
Contopus		
de Chaleco	*Contopus borealis*	Olive-sided Flycatcher
José María	*C. pertinax*	Greater Pewee
Occidental	*C. sordidulus*	Western Wood-Pewee
Verdoso	*C. virens*	Eastern Wood-Pewee
Cormorán		
Oliváceo	*Phalacrocorax olivaceus*	Olivaceous Cormorant
Orejudo	*P. auritus*	Double-crested Cormorant
Costurero		
de Agua Dulce	*Limnodromus scolopaceus*	Long-billed Dowitcher
Marino	*L. griseus*	Short-billed Dowitcher
Cuclillo		
Alirrojizo	*Coccyzus americanus*	Yellow-billed Cuckoo
Piquinegro	*C. erythropthalmus*	Black-billed Cuckoo
Ventrisucio	*C. minor*	Mangrove Cuckoo
Chipe		
Aliazul	*Vermivora pinus*	Blue-winged Warbler
Alidorado	*V. chrysoptera*	Golden-winged Warbler
Amarillo Norteño	*Dendroica petechia*	Yellow Warbler
Azul-Olivo Norteño	*Parula americana*	Northern Parula
Azul Pizarra	*Dendroica caerulescens*	Black-throated Blue Warbler
de Bachman	*Vermivora bachmanii*	Bachman's Warbler
Cabecidorado	*Protonotaria citrea*	Prothonotary Warbler
Cabecigrís Filadélfico	*Oporornis philadelphia*	Mourning Warbler
Cabecigrís Ojianillado	*O. agilis*	Connecticut Warbler
Cabecigrís de Tolmie	*O. tolmiei*	MacGillivray's Warbler
Cachetinegro	*O. formosus*	Kentucky Warbler
Carirrojo	*Cardellina rubrifrons*	Red-faced Warbler

Spanish common name	Scientific name	English common name
Celato	*Vermivora celata*	Orange-crowned Warbler
Cerúleo	*Dendroica cerulea*	Cerulean Warbler
Colifajado	*D. magnolia*	Magnolia Warbler
de Colima	*Vermivora crissalis*	Colima Warbler
de Collar	*Wilsonia canadensis*	Canada Warbler
Coronicafé	*Limnothlypis swainsonii*	Swainson's Warbler
Coroninegro	*Wilsonia pusilla*	Wilson's Warbler
Dominico	*Dendroica dominica*	Yellow-throated Warbler
Dorsinegro	*D. chrysoparia*	Golden-cheeked Warbler
Dorsiverde	*D. virens*	Black-throated Green Warbler
Encapuchado	*Wilsonia citrina*	Hooded Warbler
Galán	*Dendroica discolor*	Prairie Warbler
Gorjinaranja	*D. fusca*	Blackburnian Warbler
Gorriamarillo	*D. pensylvanica*	Chestnut-sided Warbler
Gorrinegro	*D. striata*	Blackpoll Warbler
Grupidorado	*D. coronata*	Yellow-rumped Warbler
de Kirtland	*D. kirtlandii*	Kirtland's Warbler
de Lucy	*Vermivora luciae*	Lucy's Warbler
de Nashville	*V. ruficapilla*	Nashville Warbler
Negrigrís	*Dendroica nigrescens*	Black-throated Gray Warbler
Nororiental	*D. pinus*	Pine Warbler
Occidental	*D. occidentalis*	Hermit Warbler
Pechicastaño	*D. castanea*	Bay-breasted Warbler
Peregrino	*Vermivora peregrina*	Tennessee Warbler
Pinero Gorjiamarillo	*Dendroica graciae*	Grace's Warbler
Piquigrueso	*Icteria virens*	Yellow-breasted Chat
Playero	*Dendroica palmarum*	Palm Warbler
Suelero Coronado	*Seiurus aurocapillus*	Ovenbird
Suelero Gorjiblanco	*S. motacilla*	Louisiana Waterthrush
Suelero Gorjijaspeado	*S. noveboracensis*	Northern Waterthrush
Tigrino	*Dendroica tigrina*	Cape May Warbler
de Townsend	*D. townsendi*	Townsend's Warbler
Trepador	*Mniotilta varia*	Black-and-white Warbler
Vermívoro	*Helmitheros vermivorus*	Worm-eating Warbler
de Virginia	*Vermivora virginiae*	Virginia's Warbler
Chorlito		
Alejandrino	*Charadrius alexandrinus*	Snowy Plover
Llanero	*C. montanus*	Mountain Plover
Melódico	*C. melodus*	Piping Plover
Piquigrueso	*C. wilsonia*	Wilson's Plover
Semipalmeado	*C. semipalmatus*	Semipalmated Plover
Tildío	*C. vociferus*	Killdeer
Chorlo		
Axiliclaro	*Pluvialis dominica*	Lesser Golden-Plover
Axilinegro	*P. squatarola*	Black-bellied Plover
Chotacabra		
Halcón	*Chordeiles acutipennis*	Lesser Nighthawk
Zumbón	*C. minor*	Common Nighthawk

Spanish common name	Scientific name	English common name
Chupaflor		
Coronimorado	*Eugenes fulgens*	Magnificent Hummingbird
Gorjiazul	*Lampornis clemenciae*	Blue-throated Hummingbird
Empidonax		
Alnoro	*Empidonax alnorum*	Alder Flycatcher
Canelo	*E. fulvifrons*	Buff-breasted Flycatcher
Difícil	*E. difficilis*	Pacific-slope Flycatcher
de Hammond	*E. hammondii*	Hammond's Flycatcher
Mínimo	*E. minimus*	Least Flycatcher
de Oberholser	*E. oberholseri*	Dusky Flycatcher
Occidental	*E. occidentalis*	Cordilleran Flycatcher
de Traill	*E. traillii*	Willow Flycatcher
Ventriamarillo	*E. flaviventris*	Yellow-bellied Flycatcher
Verdoso	*E. virescens*	Acadian Flycatcher
de Wright	*E. wrightii*	Gray Flycatcher
Espiza	*Spiza americana*	Dickcissel
Falaropo		
Piquifino	*Phalaropus lobatus*	Red-necked Phalarope
Piquigrueso	*P. fulicaria*	Red Phalarope
Piquilargo	*P. tricolor*	Wilson's Phalarope
Gallareta		
Americana	*Fulica americana*	American Coot
Frentirroja	*Gallinula chloropus*	Common Moorhen
Morada	*Porphyrula martinica*	Purple Gallinule
Ganso		
Cerúleo (Blanco)	*Chen caerulescens*	Snow Goose
Manchado	*Anser albifrons*	Greater White-fronted Goose
Garcita		
Oscura	*Butorides striatus*	Green-backed Heron
de Tular	*Ixobrychus exilis*	Least Bittern
Garza		
Azul	*Egretta caerulea*	Little Blue Heron
Dedos Dorados	*E. thula*	Snowy Egret
Ganadera	*Bubulcus ibis*	Cattle Egret
Nocturna Coroniclara	*Nyctanassa violacea*	Yellow-crowned Night-Heron
Nocturna Coroninegra	*Nycticorax nycticorax*	Black-crowned Night-Heron
Norteña de Tular	*Botaurus lentiginosus*	American Bittern
Piquirrosa	*Egretta rufescens*	Reddish Egret
Ventriblanca	*E. tricolor*	Tricolored Heron
Garzón		
Blanco	*Casmerodius albus*	Great Egret
Cenizo	*Ardea herodias*	Great Blue Heron
Gavilán		
Pechirrufo Mayor	*Accipiter cooperii*	Cooper's Hawk
Pechirrufo Menor	*A. striatus*	Sharp-shinned Hawk
Gaviota		
Aliglauca	*Larus glaucescens*	Glaucous-winged Gull
Apipizca	*L. pipixcan*	Franklin's Gull

Spanish common name	Scientific name	English common name
Atricila	*L. atricilla*	Laughing Gull
Californiana	*L. californicus*	California Gull
Colihendida	*Xema sabini*	Sabine's Gull
de Delaware	*Larus delawarensis*	Ring-billed Gull
Menor	*L. philadelphia*	Bonaparte's Gull
Occidental	*L. occidentalis*	Western Gull
Plateada	*L. argentatus*	Herring Gull
Gaviotina		
Común	*Sterna hirundo*	Common Tern
Dorsinegra	*S. fuscata*	Sooty Tern
de Dougall	*S. dougallii*	Roseate Tern
Elegante	*S. elegans*	Elegant Tern
de Forster	*S. forsteri*	Forster's Tern
Menor	*S. antillarum*	Least Tern
Negruzca	*Chlidonias niger*	Black Tern
Piquigruesa	*Sterna nilotica*	Gull-billed Tern
Piquinaranja	*S. maxima*	Royal Tern
Piquirroja	*S. caspia*	Caspian Tern
de Sandwich	*S. sandvicensis*	Sandwich Tern
Golondrina		
Canadiense	*Tachycineta bicolor*	Tree Swallow
Cariblanca	*T. thalassina*	Violet-green Swallow
Fulva	*Hirundo fulva*	Cave Swallow
Gorjicafé	*Stelgidopteryx serripennis*	Northern Rough-winged Swallow
Grande Negruzca	*Progne subis*	Purple Martin
Pechifajada	*Riparia riparia*	Bank Swallow
Risquera	*Hirundo pyrrhonota*	Cliff Swallow
Tijereta	*H. rustica*	Cliff Swallow
Gorrión		
Arlequín	*Chondestes grammacus*	Lark Sparrow
Bigotudo Coronirrufo	*Aimophila ruficeps*	Rufous-crowned Sparrow
Bigotudo Sonorense	*Aimophila carpalis*	Rufous-winged Sparrow
de Botteri	*A. botterii*	Botteri's Sparrow
de Brewer	*Spizella breweri*	Brewer's Sparrow
Georgiana	*Melospiza georgiana*	Swamp Sparrow
Gorriblanco	*Zonotrichia leucophrys*	White-crowned Sparrow
Indefinido Oriental	*Spizella atrogularis*	Black-chinned Sparrow
Indefinido Rayado	*S. pallida*	Clay-colored Sparrow
de Lincoln	*Melospiza lincolnii*	Lincoln's Sparrow
Llanero	*Spizella passerina*	Chipping Sparrow
Sabanero Común	*Passerculus sandwichensis*	Savannah Sparrow
Saltamonte	*Ammodramus savannarum*	Grasshopper Sparrow
Zacatero Coliblanco	*Pooecetes gramineus*	Vesper Sparrow
Grulla		
Blanca	*Grus americana*	Whooping Crane
Gris	*G. canadensis*	Sandhill Crane
Halcón		
Cernícalo	*Falco sparverius*	American Kestrel

Spanish common name	Scientific name	English common name
Esmerejón	*F. columbarius*	Merlin
Pálido	*F. mexicanus*	Prairie Falcon
Peregrino	*F. peregrinus*	Peregrine Falcon
Ibis		
Blanco	*Eudocimus albus*	White Ibis
Cara Blanca	*Plegadis chihi*	White-faced Ibis
Espátula	*Ajaia ajaja*	Roseate Spoonbill
Negro	*Plegadis falcinellus*	Glossy Ibis
Jilguero		
Canario	*Carduelis tristis*	American Goldfinch
Dorsioscuro	*C. psaltria*	Lesser Goldfinch
Lechucilla		
Colicorta	*Micrathene whitneyi*	Elf Owl
Zancón	*Speotyto cunicularia*	Burrowing Owl
Limosa		
Canela	*Limosa fedoa*	Marbled Godwit
Ornamentada	*L. haemastica*	Hudsonian Godwit
Llanero Alipálido	*Calamospiza melanocorys*	Lark Bunting
Martín Pescador Norteño	*Ceryle alcyon*	Belted Kingfisher
Mascarita Norteña	*Geothlypis trichas*	Common Yellowthroat
Milano		
Migratorio	*Ictinia mississippiensis*	Mississippi Kite
Tijereta	*Elanoides forficatus*	American Swallow-tailed Kite
Mímido		
Gris	*Dumetella carolinensis*	Gray Catbird
Pinto	*Oreoscoptes montanus*	Sage Thrasher
Mosquerito Silbador	*Camptostoma imberbe*	Northern Beardless Tyrannulet
Mosquero		
Cardenalito	*Pyrocephalus rubinus*	Vermilion Flycatcher
Fibí	*Sayornis phoebe*	Eastern Phoebe
Llanero	*S. saya*	Say's Phoebe
Ostrero Blanquinegro	*Haematopus palliatus*	American Oystercatcher
Paloma		
Aliblanca	*Zenaida asiatica*	White-winged Dove
Cabeciblanca	*Columba leucocephala*	White-crowned Pigeon
Collareja	*C. fasciata*	Band-tailed Pigeon
Común	*Zenaida macroura*	Mourning Dove
Morada Ventrioscura	*Columba flavirostris*	Red-billed Pigeon
Papamoscas		
Copetón Gorjicenizo	*M. cinerascens*	Ash-throated Flycatcher
Copetón Tiranillo	*Myiarchus tyrannulus*	Brown-crested Flycatcher
Copetón Triste	*M. tuberculifer*	Dusky-capped Flycatcher
Copetón Viajero	*M. crinitus*	Great Crested Flycatcher
Rayado Cejiblanco	*Myiodynastes luteiventris*	Sulphur-bellied Flycatcher
Patamarilla		
Mayor	*Tringa melanoleuca*	Greater Yellowlegs
Menor	*T. flavipes*	Lesser Yellowlegs
Pato		
Arcoiris	*Aix sponsa*	Wood Duck

Spanish common name	Scientific name	English common name
Boludo Menor	*Aythya affinis*	Lesser Scaup
Borrado	*A. valisineria*	Canvasback
Cabecirrojo	*A. americana*	Redhead
de Collar	*Anas platyrhynchos*	Mallard
Cucharón	*A. clypeata*	Northern Shoveler
Chalcuán	*A. americana*	American Wigeon
Friso	*A. strepera*	Gadwall
Golondrino	*A. acuta*	Northern Pintail
Mergo Copetón	*Lophodytes cucullatus*	Hooded Merganser
Mergo Pechicastaño	*Mergus serrator*	Red-breasted Merganser
Pijije Aliblanco	*Dendrocygna autumnalis*	Black-bellied Whistling-Duck
Pijije Alioscuro	*D. bicolor*	Fulvous Whistling-Duck
Piquianillado	*Aythya collaris*	Ring-necked Duck
Rojizo Alioscuro	*Oxyura jamaicensis*	Ruddy Duck
Pavito		
Aliblanco	*Myioborus pictus*	Painted Redstart
Migratorio	*Setophaga ruticilla*	American Redstart
Pelícano		
Blanco	*Pelecanus erythrorhynchos*	American White Pelican
Pardo	*P. occidentalis*	Brown Pelican
Perlita Piis	*Polioptila caerulea*	Blue-gray Gnatcatcher
Picogrueso		
Azul	*Guiraca caerulea*	Blue Grosbeak
Pechicafé	*Pheucticus melanocephalus*	Black-headed Grosbeak
Pechirrosa	*P. ludovicianus*	Rose-breasted Grosbeak
Playerito		
Alzacolita	*Actitis macularia*	Spotted Sandpiper
de Baird	*Calidris bairdii*	Baird's Sandpiper
Correlón	*C. alba*	Sanderling
Mínimo	*C. minutilla*	Least Sandpiper
Occidental	*C. mauri*	Western Sandpiper
Pradero	*Tryngites subruficollis*	Buff-breasted Sandpiper
de Rabadilla Blanca	*Calidris fuscicollis*	White-rumped Sandpiper
Semipalmeado	*C. pusilla*	Semipalmated Sandpiper
Playero		
Charquero	*Tringa solitaria*	Solitary Sandpiper
Pechirrayado	*Calidris melanotos*	Pectoral Sandpiper
Pihuihui	*Catoptrophorus semipalmatus*	Willet
Piquicorto	*Calidris canutus*	Red Knot
Roquero	*Aphriza virgata*	Surfbird
Sencillo	*Heteroscelus incanus*	Wandering Tattler
Zancón	*Calidris himantopus*	Stilt Sandpiper
Pradero		
Gorjeador	*Sturnella neglecta*	Western Meadowlark
Tortilla-con-chile	*S. magna*	Eastern Meadowlark
Ralito Negruzco	*Laterallus jamaicensis*	Black Rail
Ralo		
Barrado Grisáceo	*Porzana carolina*	Sora
Barrado Rojizo	*Rallus limicola*	Virginia Rail

Spanish common name	Scientific name	English common name
Ralón Barrado Rojizo	*Rallus elegans*	King Rail
Rascador		
Migratorio	*Pipilo chlorurus*	Green-tailed Towhee
Pinto Oscuro	*P. erythrophthalmus*	Rufous-sided Towhee
Rayador Americano	*Rynchops niger*	Black Skimmer
Reyezuelo Sencillo	*Regulus calendula*	Ruby-crowned Kinglet
Tangara		
Aliblanca Migratoria	*Piranga ludoviciana*	Western Tanager
Roja Migratoria	*P. rubra*	Summer Tanager
Roja Piquioscura	*P. flava*	Hepatic Tanager
Rojinegra Migratoria	*P. olivacea*	Scarlet Tanager
Tapacamino		
Cuerporruín	*Caprimulgus vociferus*	Whip-poor-will
de Paso	*C. carolinensis*	Chuck-will's-widow
Tevíi	*Phalaenoptilus nuttallii*	Common Poorwill
Tirano		
Dominicano	*Tyrannus dominicensis*	Gray Kingbird
Dorsinegro	*T. tyrannus*	Eastern Kingbird
Gritón	*T. vociferans*	Cassin's Kingbird
Pálido	*T. verticalis*	Western Kingbird
Piquigrueso	*T. crassirostris*	Thick-billed Kingbird
Tijereta Clara	*T. forficatus*	Scissor-tailed Flycatcher
Tropical Común	*T. melancholicus*	Tropical Kingbird
Tropical Silbador	*T. couchii*	Couch's Kingbird
Tordo		
Cabeciamarillo	*Xanthocephalus xanthocephalus*	Yellow-headed Blackbird
Cabecicafé	*Molothrus ater*	Brown-headed Cowbird
Migratorio	*Dolichonyx oryzivorus*	Bobolink
Ojiclaro	*Euphagus cyanocephalus*	Brewer's Blackbird
Ojirrojo	*Molothrus aeneus*	Bronzed Cowbird
Sargento	*Agelaius phoeniceus*	Red-winged Blackbird
Troglodita		
de Ciénega	*Cistothorus platensis*	Sedge Wren
Continental	*Troglodytes aedon*	House Wren
Pantanera	*Cistothorus palustris*	Marsh Wren
Trogón Colicobrizo	*Trogon elegans*	Elegant Trogon
Vencejito		
Alirrápido	*Chaetura vauxi*	Vaux's Swift
de Paso	*C. pelagica*	Chimney Swift
Pechiblanco	*Aeronautes saxatalis*	White-throated Swift
Vencejo Negro	*Cypseloides niger*	Black Swift
Verdugo Americano	*Lanius ludovicianus*	Loggerhead Shrike
Vireo		
Anteojillo	*Vireo solitarius*	Solitary Vireo
de Bell	*V. bellii*	Bell's Vireo
Bigotinegro	*V. altiloquus*	Black-whiskered Vireo
Filadélfico	*V. philadelphicus*	Philadelphia Vireo
Gorjeador Norteño	*V. gilvus*	Warbling Vireo

Spanish common name	Scientific name	English common name
Gorrinegro	*V. atricapillus*	Black-capped Vireo
Gris	*V. vicinior*	Gray Vireo
Ojiblanco	*V. griseus*	White-eyed Vireo
Ojirrojo Norteño	*V. olivaceus*	Red-eyed Vireo
Ojirrojo Tropical	*V. flavoviridis*	Yellow-green Vireo
Pechiamarillo	*V. flavifrons*	Yellow-throated Vireo
Vuelvepiedras Común	*Arenaria interpres*	Ruddy Turnstone
Zambullidor		
Achichilique	*Aechmophorus occidentalis*	Western Grebe
de Clark	*A. clarkii*	Clark's Grebe
Mediano	*Podiceps nigricollis*	Eared Grebe
Piquigrueso	*Podilymbus podiceps*	Pied-billed Grebe
Zarapito		
Boreal	*Numenius borealis*	Eskimo Curlew
Cabecirrayado	*N. phaeopus*	Whimbrel
Ganga	*Bartramia longicauda*	Upland Sandpiper
Piquilargo	*Numenius americanus*	Long-billed Curlew
Zorzal Pechirrojo	*Turdus migratorius*	American Robin
Zorzalito		
Carigrís	*Catharus minimus*	Gray-cheeked Thrush
Colirrufo	*C. guttatus*	Hermit Thrush
Maculado	*C. mustelinus*	Wood Thrush
de Swainson	*C. ustulatus*	Swainson's Thrush
de Wilson	*C. fuscescens*	Veery

LITERATURE CITED

Abbott, D. J., III, and D. W. Finch. 1978. First Variegated Flycatcher (*Empidonomus varius*) record for the United States. *American Birds* 32:161–163.

Able, K. P. 1973. The role of weather variables and flight direction in determining the magnitude of nocturnal bird migration. *Ecology* 54:1031–1041.

Abramson, I. J. 1976. The Black Hawk in South Florida. *American Birds* 30:661–662.

Albuquerque, J. L. B. 1978. Contribution to the knowledge of *Falco peregrinus* in South America. *Revista Brasiliera Biológica* 38:727–737.

Aldrich, J. W., and R. W. Coffin. 1980. Breeding bird populations from forest to suburbia after 37 years. *American Birds* 34:3–7.

Aldrich, J. W., and C. S. Robbins. 1970. Changing abundance of migratory birds in North America. In *A symposium of the Smithsonian Institution on the avifauna of northern Latin America,* ed. H. K. Buechner and J. H. Buechner, 17–26. Smithsonian Contributions to Zoology, no. 26. Washington, D.C.

Ali, S., and S. D. Ripley. 1968–1974. *Handbook of the birds of India and Pakistan.* 10 vols. Oxford University Press, Madras, India.

Allen, J. A. 1880. Origin of the instinct of migration in birds. *Bulletin of the Nuttall Ornithological Club* 5:151–154.

Alverson, W. S., D. M. Waller, and S. L. Solheim. 1988. Forests too deer: Edge effects in northern Wisconsin. *Conservation Biology* 2:348–358.

Ambuel, B., and S. A. Temple. 1982. Songbird populations in southern Wisconsin forests: 1954 and 1979. *Journal of Field Ornithology* 53:149–158.

———. 1983. Area-dependent changes in the bird communities and vegetation of southern Wisconsin forests. *Ecology* 64:1057–1068.

American Ornithologists' Union. 1976. Thirty-third supplement to the American Ornithologists' Union *Check-list of North American birds. Auk* 93:875–879.

———. 1983. *Check-list of North American birds.* 6th ed. American Ornithologists' Union, Washington, D.C.

———. 1985. Thirty-fifth supplement to the American Ornithologists' Union *Check-list of North American birds. Auk* 102:680–686.

———. 1987. Thirty-sixth supplement to the American Ornithologists' Union *Check-list of North American birds. Auk* 104:591–596.

———. 1989. Thirty-seventh supplement to the American Ornithologists' Union *Check-list of North American birds. Auk* 106:532–538.

217

————. 1991. Thirty-eighth supplement to the American Ornithologists' Union *Check-list of North American birds. Auk* 108:750–754.

————. 1993. Thirty-ninth supplement to the American Ornithologists' Union *Check-list of North American birds. Auk* 110:675–682.

Andrade, A. 1975. The chance that Mexico will be a desert in 25 years [in Spanish]. *Mexican Forestry* 49:12–13.

Andrade, A., and V. Payan. 1973. The country in risk of being converted to a barren plain [in Spanish]. *Mexican Forestry* 47:31–32.

Andrle, R. F. 1964. A biogeographical investigation of the Sierra de Tuxtla in Veracruz, Mexico. Ph.D. dissertation, Louisiana State University, Baton Rouge.

————. 1967. Birds of the Sierra de Tuxtla in Veracruz, Mexico. *Wilson Bulletin* 79:163–187.

————. 1968. Raptors and other North American migrants in Mexico. *Condor* 70:393–395.

Arendt, W. J., with collaborators. 1992. Status of North American migrant landbirds in the Caribbean region: A summary. In *Ecology and conservation of Neotropical migrant landbirds,* ed. J. M. Hagan III and D. W. Johnston, 143–174. Smithsonian Institution Press, Washington, D.C.

Armbruster, M. J., T. S. Baskett, W. R. Goforth, and S. K. Sadler. 1978. Evaluating call-count procedures for measuring local Mourning Dove populations. *Transactions of the Missouri Academy of Sciences* 12:75–90.

Armitage, K. 1955. Territorial behavior in fall migrant Rufous Hummingbirds. *Condor* 57:239–240.

Arnold, K. 1981. Fidelity of Common Snipe to wintering grounds, with comments on local movements. *Southwestern Naturalist* 26:319–321.

Arrenhius, O. 1921. Species and area. *Journal of Ecology* 9:95–99.

Askins, R. A., and M. J. Philbrick. 1987. Effects of changes in regional forest abundance on the decline and recovery of a forest bird community. *Wilson Bulletin* 99:7–21.

Askins, R. A., M. J. Philbrick, and D. S. Sugeno. 1987. Relationship between the regional abundance of forest and the composition of forest bird communities. *Biological Conservation* 39:129–152.

Askins, R. A., J. F. Lynch, and R. Greenberg. 1990. Population declines in migratory birds in eastern North America. *Current Ornithology* 7:1–57.

Askins, R. A., D. N. Ewert, and R. L. Norton. 1992. Abundance of wintering migrants in fragmented and continuous forests in the U.S. Virgin Islands. In *Ecology and conservation of Neotropical migrant landbirds,* ed. J. M. Hagan III and D. W. Johnston, 197–206. Smithsonian Institution Press, Washington, D.C.

Audubon, J. J. 1840–1844. *The birds of America.* 7 vols. Audubon and Chevalier, New York.

Austin, O. L., Jr. 1985. *Families of birds.* 3d ed. Golden Press, New York.

Ayers, A. Y., P. W. Sykes, Jr., and W. J. Sykes. 1980. Two more records of the Tropical Kingbird for Florida. *Florida Field Naturalist* 8:25–26.

Azara, F. de. 1802–1805. *Notes on the natural history of the birds of Paraguay and the Rio de la Plata* [in Spanish]. 3 vols. Madrid.

Baez, A. P., R. de Nulman, I. Rosas, and L. C. Galvez. 1976. Aquatic organism contamination by mercury residues in the Coatzacoalcos River estuary, Mexico. In *Trace contaminants of agriculture, fisheries, and food in developing countries,* 73–79.

Bailey, R. O. 1979. Methods of estimating total lipid content in the Redhead duck (*Aythya americana*) and an evaluation of condition indices. *Canadian Journal of Zoology* 57:1830–1833.

Bain, G. A. C. 1980. The relationship between preferred habitat, physical condition, and hunting mortality of Canvasbacks (*Aythya valisineria*) and Redheads (*A. americana*) at Long Point, Ontario. Master's thesis, University of Western Ontario, London, Ont.

Baird, T. H. 1990. *Changes in breeding bird populations between 1930 and 1985 in the Quaker Run Valley of Allegheny State Park, New York.* New York State Museum Bulletin no. 477. State University of New York at Albany.

Bairlein, F. 1983. Habitat selection and associations of species in European passerine birds during southward, post-breeding migrations. *Ornis Scandinavica* 14:239–245.

———. 1985. Body weights and fat deposition of Palaearctic passerine migrants in the central Sahara. *Oecologia* 66:141–146.

———. 1988. How do migratory songbirds cross the Sahara? *Trends in Ecology and Evolution* 3:191–194.

———. 1990. Nutrition and food selection in migratory birds. In *Bird migration: Physiology and ecophysiology,* ed. E. Gwinner, 198–213. Springer-Verlag, Berlin.

———. 1992. Morphology-habitat relationships in migrating songbirds. In *Ecology and conservation of Neotropical migrant landbirds,* ed. J. M. Hagan III and D. W. Johnston, 356–369. Smithsonian Institution Press, Washington, D.C.

Baker, M. C., and A. E. M. Baker. 1973. Niche relationships among six species of shorebirds on their wintering and breeding ranges. *Ecological Monographs* 43: 193–212.

Baker, R. E., and G. A. Parker. 1979. The evolution of bird coloration. *Philosophical Transactions of the Royal Society, London,* ser. B, 287:63–130.

Baker, R. R. 1978. *The evolutionary ecology of animal migration.* Holmes and Meier, New York.

Barducci, T. B. 1972. Ecological consequences of pesticides used for the control of cotton insects in Cañete Valley, Peru. In *The careless technology: Ecology and international development,* 423–438.

Barlow, J. C. 1980. Patterns of ecological interactions among migrant and resident vireos on the wintering grounds. In *Migrant birds in the Neotropics: Ecology, behavior, distribution, and conservation,* ed. A. Keast and E. S. Morton, 79–107. Smithsonian Institution Press, Washington, D.C.

Barnett, S. A. 1958. Experiments in "neophobia" in wild and laboratory rats. *British Journal of Psychology* 49:195–201.

Barrows, W. B. 1883. Birds of the lower Uruguay. *Bulletin of the Nuttall Ornithological Club* 8:82–94, 128–143, 198–212.

Bartgis, R. 1992. Loggerhead Shrike (*Lanius ludovicianus*). In *Migratory nongame birds of management concern in the northeast,* ed. K. J. Schneider and D. M. Pence, 281–297. U.S. Fish and Wildlife Service, Region 5, Newton Corner, Mass.

Bartinicki, E. A. 1979. Inca Dove in Comanche County, Oklahoma. *Bulletin of the Oklahoma Ornithological Society* 12:31–32.

Bartram, W. 1791. *Travels through North and South Carolina, Georgia, east and west Florida.* Jones and Johnson, Philadelphia.

Baskett, T. S., M. J. Armbruster, and M. W. Sayre. 1978. Biological perspectives for the Mourning Dove call-count survey. *Transactions of the North American Wildlife and Natural Resources Conference* 37:312–325.

Bednarz, J. C., D. Klem, Jr., L. J. Goodrich, and S. E. Senner. 1990. Migration counts of raptors at Hawk Mountain, Pennsylvania, as indicators of population trends, 1934–1986. *Auk* 107:96–109.

Beebe, W. 1947. Avian migration at Rancho Grande in north-central Venezuela. *Zoologica* 32:153–168.

Bellrose, F. C. 1976. *Ducks, geese, and swans of North America.* Stackpole Books, Harrisburg, Pa.

Belton, W. 1984. Birds of Rio Grande do Sul, Brazil. Part 1. *Bulletin of the American Museum of Natural History* 178:369–631.

———. 1985. Birds of Rio Grande do Sul, Brazil. Part 2. *Bulletin of the American Museum of Natural History* 180:1–241.

Bennett, S. E. 1980. Interspecific competition and the niche of the American Redstart (*Setophaga ruticilla*) in wintering and breeding communities. In *Migrant birds in the Neotropics: Ecology, behavior, distribution, and conservation,* ed. A. Keast and E. S. Morton, 319–335. Smithsonian Institution Press, Washington, D.C.

Bent, A. C. 1953. *Life histories of North American wood warblers.* Bulletin of the U.S. National Museum no. 203. Washington, D.C.

Bergman, D. 1994. The ecology of the Northern Pintail on its wintering grounds in south Texas. Master's thesis, Texas A&I University, Kingsville.

Bermingham, E., S. Rohwer, S. Freeman, and C. Wood. 1992. Vicariance biogeography in the Pleistocene and speciation in North American wood warblers: A test of Mengel's model. *Proceedings of the National Academy of Sciences* (USA) 89: 6624–6628.

Bernal, I. 1969. *The Olmec world.* University of California Press, Berkeley and Los Angeles.

Berthold, P. 1975. Migration: Control and metabolic physiology. In *Avian biology,* vol. 5, ed. D. S. Farner and J. R. King, 77–128. Academic Press, New York.

———. 1988. The control of migration in European warblers. *Proceedings of the International Ornithological Congress* 19:215–249.

———. 1993. *Bird migration: A general survey.* Oxford University Press, Oxford.

Berthold, P., F. Bairlein, and U. Querner. 1976. Uber die Verteilung von ziehenden

Kleinvögeln in Rastbiotopen und den Fangerfolg von Fanganlagen. *Vogelwarte* 28:267–273.

Bibby, C. F., and R. E. Green. 1980. Foraging behavior of migrant Pied Flycatchers, *Ficedula hypoleuca,* on temporary territories. *Journal of Animal Ecology* 49: 507–521.

Birch, T. W., and E. H. Wharton. 1982. Land-use change in Ohio, 1952–1979. U.S. Department of Agriculture, Forest Service Resource Bulletin no. NE-70.

Blake, J. G., and B. A. Loiselle. 1992. Habitat use by Neotropical migrants at La Selva Biological Station and Braulio Carrillo National Park, Costa Rica. In *Ecology and conservation of Neotropical migrant landbirds,* ed. J. M. Hagan III and D. W. Johnston, 257–272. Smithsonian Institution Press, Washington, D.C.

Blake, J. G., G. E. Niemi, and J. A. Hanowski. 1992. Drought and annual variation in bird populations: Effects of migratory strategy and breeding habitat. In *Ecology and conservation of Neotropical migrant landbirds,* ed. J. M. Hagan III and D. W. Johnston, 419–430. Smithsonian Institution Press, Washington, D.C.

Blankinship, D. R., J. G. Teer, W. H. Kiel, Jr., and J. B. Trefethen. 1972. Movements and mortality of White-winged Doves banded in Tamaulipas, Mexico. *Transactions of the North American Wildlife and Natural Resources Conference* 37:312–325.

Bohlen, H. D. 1976. A Great-tailed Grackle from Illinois. *American Birds* 30:917.

Bollinger, E. K., and T. A. Gavin. 1992. Eastern Bobolink populations: Ecology and conservation in an agricultural landscape. In *Ecology and conservation of Neotropical migrant landbirds,* ed. J. M. Hagan III and D. W. Johnston, 497–506. Smithsonian Institution Press, Washington, D.C.

Bond, J. 1957. *Second supplement to the check-list of birds of the West Indies.* Academy of Natural Sciences of Philadelphia.

Boyd, R. L. 1978. Rivoli's Hummingbird in Linn County, Kansas. *Bulletin of the Kansas Ornithological Society* 29:10–11.

Boyle Lemus, W. A., C. M. Corletto Tobar, and R. U. Oliva Aquilar. 1976. Evaluation of toxic metal contamination of superficial waters and sediments of the hydrographic basin of the Acelhuate River and metropolitan zone of San Salvador [in Spanish]. Universidad de El Salvador, San Salvador.

Bradshaw, D. 1992. Virginia joins the ranks of Partners in Flight. *Virginia Wildlife* (May): 1–15.

Branquinho, C. L., and V. J. Robinson. 1976. Some aspects of lead pollution in drinking water and air in Rio de Janeiro. *Environmental Pollution* 10:287–292.

Brera, A. M., and F. Shahrokhi. 1978. Application of Landsat data to monitor desert spreading in the Sahara region. *Proceedings of Remote Sensing of the Environment Society* 12:1329–1335.

Briggs, R. L. 1977. Blue-winged Teal banding project Panama Canal Zone. *North American Bird Bander* 2:104–105.

Briggs, S. A., and J. H. Criswell. 1979. Gradual silencing of spring in Washington. *Atlantic Naturalist* 32:19–26.

Briinig, E. F. 1977. The tropical rain forest—A wasted asset or an essential biospheric resource? *Ambio* 6:187–191.

Brittingham, M. C., and S. A. Temple. 1983. Have cowbirds caused forest songbirds to decline? *Bioscience* 33:31–35.

Brodkorb, P. 1948. Some birds from the lowlands of central Veracruz, Mexico. *Quarterly Journal of the Florida Academy of Sciences* 10:31–38.

Brosset, A. 1968. Ecological localization of migratory birds in the equatorial forest of Gabon [in French]. *Biologica Gabonica* 4:211–226.

———. 1984. The ecological place of migratory birds in the equatorial forest of Gabon [in French]. *Alauda* 52:81–101.

Brown, J. L. 1964. The evolution of diversity in avian territorial systems. *Wilson Bulletin* 76:160–169.

———. 1969. Territorial behavior and population regulation in birds. *Wilson Bulletin* 81:293–329.

Brown, L. H., E. K. Urban, and K. Newman. 1982. *The birds of Africa.* Vol. 1. Academic Press, New York.

Buechner, H. K., and J. H. Buechner, eds. 1970. *A symposium of the Smithsonian Institution on the avifauna of northern Latin America.* Smithsonian Contributions to Zoology, no. 26. Washington, D.C.

Bull, J. L. 1974. *Birds of New York State.* Doubleday, Garden City, N.Y.

Burger, J., M. Fitch, G. Shugart, and W. Werther. 1980. Piracy in *Larus* gulls at a dump in New Jersey. *Proceedings of the Colonial Waterbird Group* 3:87–98.

Buskirk, R. E., and W. H. Buskirk. 1976. Changes in arthropod abundance in a highland Costa Rican forest. *American Midland Naturalist* 95:288–298.

Buskirk, W. H. 1972. Foraging ecology of birds in a tropical forest. Ph.D. dissertation, University of California, Davis.

———. 1980. Influence of meteorological patterns and trans-Gulf migration on the calendars of latitudinal migrants. In *Migrant birds in the Neotropics: Ecology, behavior, distribution, and conservation,* ed. A. Keast and E. S. Morton, 485–491. Smithsonian Institution Press, Washington, D.C.

Buskirk, W. H., G. V. N. Powell, J. F. Wittenberger, R. E. Buskirk, and T. U. Powell. 1972. Interspecific bird flocks in tropical highland Panama. *Auk* 89:612–624.

Butcher, G. S., W. A. Niering, W. J. Barry, and R. H. Goodwin. 1981. Equilibrium biogeography and the size of nature preserves: An avian case study. *Oecologia* 49: 29–37.

Butcher, G. S., and S. Rohwer. 1989. The evolution of conspicuous and distinctive coloration for communication in birds. *Current Ornithology* 6:51–108.

Cain, S. A., and G. M. deO. Castro. 1959. *Manual of vegetation analysis.* Harper and Row, New York.

Calvert, W. H., L. E. Hedrick, and L. P. Brower. 1979. Mortality of the monarch butterfly (*Danaus plexippus*): Avian predation at five overwintering sites in Mexico. *Science* 204:847–851.

Canestri, V., and O. Ruiz. 1973. The destruction of mangroves. *Marine Pollution Bulletin* 4:183–185.

Capurro, H. A., and E. H. Bucher. 1988. Annotated list of the birds of the chacoan forest at Joaquín V. González, Salta, Argentina [in Spanish]. *Hornero* 13:39–46.

Carson, R. 1962. *Silent spring.* Houghton Mifflin, New York.

Carter, J. W. 1992. Upland Sandpiper (*Bartramia longicauda*). In *Migratory nongame birds of management concern in the northeast,* ed. K. J. Schneider and D. M. Pence, 235–251. U.S. Fish and Wildlife Service, Region 5, Newton Corner, Mass.

Caufield, C. 1984. *In the rainforest.* Knopf, New York.

Chapin, J. P. 1932. The birds of the Belgian Congo: Part 1. *Bulletin of the American Museum of Natural History,* vol. 65.

Chesser, R. T. 1994. Migration in South America: An overview of the austral system. *Bird Conservation International* 4:91–107.

Chipley, R. M. 1976. The impact of wintering migrant Wood Warblers on resident insectivorous passerines in a subtropical Colombian oak woods. *Living Bird* 15: 119–141.

Ciflentes Lemus, J. L., G. L. Kesteven, A. Zarur, and A. Medina. 1971. Aspects of contamination of sea water in the Gulf of Mexico. *FAO Fish Report* (Food and Agricultural Organization, United Nations) 99:112.

Clark, D. R., Jr., and A. J. Krynitsky. 1983. DDT: Recent contamination in New Mexico and Arizona? *Environment* 25:27–31.

Cody, M. L. 1974. *Competition and the structure of bird communities.* Princeton University Press, Princeton, N.J.

Coffey, B. B., Jr. 1943. Post-juvenal migration of herons. *Bird-Banding* 14:34–39.

———. 1948. Southward migration of herons. *Bird-Banding* 19:1–5.

Cohen, D. 1967. Optimization of seasonal migratory behavior. *American Naturalist* 101:5–18.

Coiner, J. C. 1980. Using Landsat to monitor changes in vegetation cover induced by desertification processes. *Proceedings of Remote Sensing of the Environment Society* 14:1341–1347.

Colston, P. R., and K. Curry-Lindahl. 1986. *The birds of Mount Nimba, Liberia.* British Museum of Natural History, London.

Companhía Técnica Saneamento Ambiental. 1978. *Subterranean water pollution in the state of São Paulo* [in Portuguese]. Companhía Técnica Saneamento Ambiental, São Paulo, Brazil.

Confer, J. L. 1992. Golden-winged Warbler (*Vermivora chrysoptera*). In *Migratory nongame birds of management concern in the northeast,* ed. K. J. Schneider and D. M. Pence, 369–383. U.S. Fish and Wildlife Service, Region 5, Newton Corner, Mass.

Conner, R. N., J. G. Dickson, and J. H. Williamson. 1983. A comparison of breeding bird census techniques with mist netting results. *Wilson Bulletin* 95:276–280.

Cooke, M. T. 1938. Returns of banded birds: Recoveries of banded marsh birds. *Bird-Banding* 9:80–87.

———. 1946. The winter range of the Great Blue Heron. *Auk* 63:254.

———. 1950. Returns from banded birds. *Bird-Banding* 21:11–17.

Cooke, W. W. 1915. *Bird migration.* U.S. Department of Agriculture Bulletin no. 185. Washington, D.C.

Cox, G. W. 1968. The role of competition in the evolution of migration. *Evolution* 22:180–192.

———. 1985. The evolution of avian migration systems between temperate and tropical regions of the New World. *American Naturalist* 126:451–474.

Criswell, J. H. 1975. Breeding bird population studies, 1975. *Atlantic Naturalist* 30:175–176.

Cruden, R. W., and V. M. Toledo. 1976. Oriole pollination of *Erythrina breviflora* (Leguminosae): Evidence for a polytypic view of ornithophily. *Plant Systematics and Evolution* 126:393–403.

Cruz, A. 1972. Birds of the Lluidas Vale (Worthy Park) Region, Jamaica. *Quarterly Journal of the Florida Academy of Sciences* 35:72–80.

———. 1974. Feeding assemblages of Jamaican birds. *Condor* 76:103–107.

Danforth, S. T. 1937. Ornithological investigations in Vieques Island, Puerto Rico, during December, 1935. *Journal of the Department of Agriculture, Puerto Rico* 21:539–550.

———. 1939. The birds of Montserrat. *Journal of the Department of Agriculture, Puerto Rico* 23:47–66.

Darwin, C. 1871. *The descent of man.* Random House, Modern Library, New York.

Davis, S. H. 1977. *Victims of the miracle: Development and the Indians of Brazil.* Cambridge University Press, New York.

De Goody, M. P. 1971. Pollution of the Moji-Guacu River. Part 1. Problems caused by the bacteria *Sphaerotilus natans. Ciencias de Cultivación* (São Paulo) 23: 199–204.

DeGraaf, R., and J. H. Rappole. 1995. *Neotropical migratory birds.* Cornell University Press, Ithaca, N.Y. In press.

Delap, E. 1979. Groove-billed Ani in Washington County, Oklahoma. *Bulletin of the Oklahoma Ornithological Society* 12:32–33.

De Ploey, J., and O. Cruz. 1979. Landslides in the Serra Do Mar, Brazil. *Catena* 6:111–122.

de Roo, A., and J. Deheeger. 1969. Ecology of the Great Reed Warbler (*Acrocephalus arundinaceus*) (L.) wintering in the southern Congo savanna. *Gerfaut* 59:260–275.

DesGranges, J. L., and P. R. Grant. 1980. Migrant hummingbirds' accommodation into tropical communities. In *Migrant birds in the Neotropics: Ecology, behavior, distribution, and conservation,* ed. A. Keast and E. S. Morton, 395–409. Smithsonian Institution Press, Washington, D.C.

Diamond, A. W., and R. W. Smith. 1973. Returns and survival of banded warblers wintering in Jamaica. *Bird-Banding* 44:221–224.

Dianese, J. C., P. Pigati, and K. Kitayama. 1976. Residues of chlorinated hydrocarbon pesticides in the Lake of Brasilia. *Biología* 42:151–155.

Dick, J. H. 1974. A new species for South Carolina: Fork-tailed Flycatcher photographed on Bull's Island. *Chat* 38:73–74.

Dickey, D. R., and A. J. van Rossem. 1938. *The birds of El Salvador.* Field Museum of Natural History, Zoological Series, no. 23.

Dickerman, R. W. 1971. Further notes on Costa Rican birds. *Condor* 73:252–253.

Dilger, W. C. 1956a. Adaptive modifications and ecological isolating mechanisms in the thrush genera *Catharus* and *Hylocichla. Wilson Bulletin* 68:171–199.

———. 1956b. Hostile behavior and reproductive isolating mechanisms in the avian genera *Catharus* and *Hylocichla. Auk* 73:314–353.

Dirzo, R., and M. C. Garcia. 1992. Rates of deforestation in Los Tuxtlas, a Neotropical area in southeast Mexico. *Conservation Biology* 6:84–90.

Dixon, C. 1897. *The migration of birds.* Horace Cox, Windsor House, London.

Droege, S. 1991. Unpublished summary of Breeding Bird Survey data, 1966–1989, in letter to Neotropical Migrant Workshop participants. Office of Migratory Bird Management, U.S. Fish and Wildlife Service, Laurel, Md.

Drury, W. H., and J. A. Keith. 1962. Radar studies of songbird migration in coastal New England. *Ibis* 104:449–489.

Dufour, K. W., C. D. Ankney, and P. J. Weatherhead. 1993. Condition and vulnerability to hunting among mallards staging at Lake St. Clair, Ontario. *Journal of Wildlife Management* 57:209–215.

Duran Bernales, F. 1970. *Soil conservation and erosion: A dying land* [in Spanish]. Zig-Zag, Santiago, Chile.

Eaton, E. H. 1910. *Birds of New York.* State University of New York at Albany.

Eaton, S. W. 1953. Wood warblers wintering in Cuba. *Wilson Bulletin* 65:169–174.

Edwards, E. P. 1972. *A field guide to the birds of Mexico.* E. P. Edwards, Sweet Briar, Va.

Edwards, E. P., and R. E. Tashian. 1959. Avifauna of the Catemaco basin of southern Veracruz, Mexico. *Condor* 61:325–337.

Eisenmann, E., and H. Loftin. 1968. Birds of the Panama Canal Zone area, Panama (1967). *Florida Naturalist* 41:57–60, 95.

———. 1971. *Field check-list of birds of the Panama Canal Zone area.* 2d ed. Florida Audubon Society, Maitland, Fla.

Elgood, J. R., R. E. Sharland, and P. Ward. 1966. Palaearctic migrants in Nigeria. *Ibis* 108:84–116.

Ely, C. A. 1973. Returns and recoveries of North American birds banded in southern Mexico. *Bird-Banding* 44:228–229.

Ely, C. A., P. J. Latas, and R. R. Lohoefener. 1977. Additional returns and recoveries of North American birds banded in southern Mexico. *Bird-Banding* 48:275–276.

Emlen, J. T. 1973. Territorial aggression in wintering warblers at Bahama agave blossoms. *Wilson Bulletin* 85:71–74.

———. 1977. The land bird populations of Grand Bahama Island: A study in faunal, community, and niche dynamics. *Ornithological Monographs,* no. 24.

Emlen, S. T. 1975. Migration, orientation, and navigation. In *Avian biology,* vol. 5, ed. D. S. Farner and J. R. King, 129–219. Academic Press, New York.

Erskine, A. J., B. T. Collins, E. Hayakawa, and C. Downes. 1992. The cooperative Breeding Bird Survey in Canada, 1989–1991. *Canadian Wildlife Service, Program Notes* 199:1–14.

Faaborg, J., and J. E. Winters. 1979. Winter resident returns and longevity and weights of Puerto Rican birds. *Bird-Banding* 50:216–223.

Farner, D. S. 1955. The annual stimulus for migration. In *Recent studies in avian biology,* ed. A. Wolfson, 198–237. University of Illinois Press, Champaign.

Farnsworth, E. G., and F. G. Golley, eds. 1974. *Fragile ecosystems: Evolution of research and application in the Neotropics.* Springer-Verlag, New York.

Feinsinger, P. 1980. Asynchronous migration patterns and the coexistence of tropical hummingbirds. In *Migrant birds in the Neotropics: Ecology, behavior, distribution, and conservation,* ed. A. Keast and E. S. Morton, 411–419. Smithsonian Institution Press, Washington, D.C.

Fernandez, A. E. 1974. Some observations on the contamination of the coastal waters of the city of Cumaná. *Boletín del Instituto Oceanográfico de la Universidad Oriente Cumaná* 12:23–32.

Ficken, M. S., and R. W. Ficken. 1962. The comparative ethology of wood warblers: A review. *Living Bird* 1:103–118.

———. 1967. Age-specific differences in the breeding behavior of the American Redstart. *Wilson Bulletin* 79:188–199.

Figueroa, L., P. Torres, R. P. Schlater, F. Asenjo, R. Franjola, and B. Contreras. 1979. Research on Pseudophylloidea from the south of Chile. 3. Investigation of diphyllobothrium-SP from birds of Calafquen Lake. *Boletín Chiliano de Parasitología* 34:13–20.

Fitzpatrick, J. W. 1980. Wintering of North American tyrant flycatchers in the Neotropics. In *Migrant birds in the Neotropics: Ecology, behavior, distribution, and conservation,* ed. A. Keast and E. S. Morton, 67–78. Smithsonian Institution Press, Washington, D.C.

Fjeldså, J., and N. Krabbe. 1990. *Birds of the high Andes.* Zoology Museum, University of Copenhagen, Copenhagen.

Fogden, M. P. L. 1972. The seasonality and population dynamics of equatorial forest birds in Sarawak. *Ibis* 114:307–343.

Foster, G. M. 1942. A primitive Mexican economy. *Monographs of the American Ethnological Society* 5:1–115.

Frankel, A. L., and T. S. Baskett. 1961. The effect of pairing on cooing of penned Mourning Doves. *Journal of Wildlife Management* 25:372–384.

Freemark, K. E., and B. Collins. 1992. Landscape ecology of birds breeding in temperate forest fragments. In *Ecology and conservation of Neotropical migrant land-*

birds, ed. J. M. Hagan III and D. W. Johnston, 443–454. Smithsonian Institution Press, Washington, D.C.

Freese, F. 1979. Fork-tailed Flycatcher in Columbia County (Wisconsin). *Passenger Pigeon* 41:41–42.

Fretwell, S. D. 1969. Dominance behavior and winter habitat distribution in juncos (*Junco hyemalis*). *Bird-Banding* 34:293–306.

———. 1972. *Populations in a seasonal environment.* Princeton University Press, Princeton, N.J.

Fretwell, S. D., and H. L. Lucas. 1970. On territorial behavior and other factors influencing habitat distribution in birds. Part 1. Theoretical development. *Acta Biotheoretica* 19:16–36.

Friedmann, H., and F. D. Smith, Jr. 1950. A contribution of the ornithology of northeastern Venezuela. *Proceedings of the U.S. National Museum* 100:411–538.

Fuentes Godo, P. 1974. Soil erosion in Plata Valley, Brazil. *Ciencias de Investigación* 30:298–302.

Galli, A. E., C. F. Leck, and R. T. T. Forman. 1976. Avian distribution patterns in forest islands of different sizes in central New Jersey. *Auk* 93:356–364.

Gauthreaux, S. A., Jr. 1978. The influence of global climatological factors on the evolution of bird migratory pathways. *Proceedings of the International Ornithological Congress* 17:517–525.

———. 1982. The ecology and evolution of avian migration systems. In *Avian biology,* vol. 6, ed. D. S. Farner, J. R. King, and K. C. Parkes, 93–168. Academic Press, New York.

Gehlbach, F. R., D. O. Dillon, H. L. Harell, S. E. Kennedy, and K. R. Wilson. 1976. Avifauna of the Rio Corona, Tamaulipas, Mexico: Northeastern limit of the tropics. *Auk* 93:53–65.

Gladstone, D. E. 1983. *Bubulcus ibis* (Garcilla Bueyera, Cattle Egret). In *Costa Rican natural history,* ed. D. Janzen, 550–551. University of Chicago Press, Chicago.

Gleason, H. A. 1922. On the relationship between species and area. *Ecology* 3:158–162.

Gochfeld, M. 1980. Mercury levels in some sea birds of the Humboldt Current, Peru. *Environmental Pollution* 22:197–206.

Goldwasser, S., D. Gaines, and S. R. Wilbur. 1980. The Least Bell's Vireo in California: A de facto endangered race. *American Birds* 34:742–745.

Gomez-Pompa, A., C. Vazquez-Yanes, and S. Guevara. 1972. The tropical rain forest: A non-renewable resource. *Science* 177:762–765.

Goodland, R. J. A., and H. S. Irwin. 1974. An ecological discussion of the environmental impact of the highway construction program in the Amazon basin. *Landscape Planning* 1:123–254.

Gore, M. E. J., and A. R. M. Gepp. 1978. *Birds of Uruguay* [in Spanish]. Mosca Hermanos, Montevideo, Uruguay.

Gorski, L. J. 1969. Traill's Flycatchers of the "fitz-bew" songform wintering in Panama. *Auk* 86:745–747.

————. 1971. Traill's Flycatchers of the Fee-Bee-O songform wintering in Peru. *Auk* 88:429–431.

Goss-Custard, J. D. 1970. Feeding dispersion in some over-wintering wading birds. In *Social behavior in birds and mammals,* ed. J. H. Crook, 3–35. Academic Press, London.

Gradwohl, J., and R. Greenberg. 1980. The formation of antwren flocks on Barro Colorado Island, Panama. *Auk* 97:385–395.

Grainger, A. 1980. The state of the world's tropical forests. *Ecologist* 10:6–19.

Gräser, K. 1905. *Der Zug der Vögel.* H. Walther, Berlin.

Green, K. M., J. F. Lynch, J. Sircar, and L. S. Z. Greenberg. 1987. LANDSAT remote sensing to assess habitat for migratory birds in the Yucatán Peninsula, Mexico. *Vida Silvestre* 1:27–38.

Greenberg, R. 1979. Body size, breeding habitat, and winter exploitation systems in *Dendroica. Auk* 96:756–766.

————. 1980. Demographic aspects of long-distance migration. In *Migrant birds in the Neotropics: Ecology, behavior, distribution, and conservation,* ed. A. Keast and E. S. Morton, 493–504. Smithsonian Institution Press, Washington, D.C.

————. 1981a. Dissimilar bill shapes in New World tropical versus temperate forest foliage-gleaning birds. *Oecologia* 49:143–147.

————. 1981b. Frugivory in some migrant tropical forest wood warblers. *Biotropica* 13:215–223.

————. 1983. The role of neophobia in determining the degree of foraging specialization in some migrant warblers. *American Naturalist* 122:444–453.

————. 1984a. Differences in feeding neophobia in the tropical migrant wood warblers *Dendroica castanea* and *D. pensylvanica. Journal of Comparative Psychology* 98:131–136.

————. 1984b. Neophobia in the foraging site selection of a Neotropical migrant bird: An experimental study. *Proceedings of the National Academy of Sciences* (USA) 81:3778–3780.

————. 1984c. *The winter exploitation system of Bay-breasted and Chestnut-sided warblers in Panama.* University of California Publications in Zoology, vol. 116. Berkeley.

————. 1986. Competition in migrant birds in the nonbreeding season. *Current Ornithology* 3:281–307.

————. 1987a. Development of dead leaf foraging in a tropical migrant warbler. *Ecology* 68:130– 141.

————. 1987b. Seasonal foraging specialization in the Worm-eating Warbler. *Condor* 89:158–168.

————. 1990. Ecological plasticity, neophobia, and resource use in birds. *Studies in Avian Biology* 13:431–437.

Greenberg, R., and J. A. Gradwohl. 1980. Observations of paired Canada Warblers *Wilsonia canadensis* during migration in Panama. *Ibis* 122:509–512.

Greenberg, R., and J. Salgado Ortiz. 1994. Interspecific defense of pasture trees by wintering Yellow Warblers. *Auk* 111:672–682.

Greenwood, P. J. 1980. Mating systems, philopatry, and dispersal in birds and mammals. *Animal Behaviour* 28:1140–1162.

Gregersen, H. N., and A. Contreras. 1975. *U.S. investment in the forest-based sector in Latin America.* Johns Hopkins University Press, Baltimore.

Groebbels, F. 1928. Zur Physiologie des Vogelzuges. *Verhandlungen der Ornithologischen Gesellschaft in Bayern* 18:44–74.

Gwinner, E. 1986. Circannual rhythms in the control of avian migrations. *Advanced Studies in Behavior* 16:191–228.

———. 1990. Circannual rhythms in bird migration: Control of temporal patterns and interactions with photoperiod. In *Bird migration: Physiology and ecophysiology,* ed. E. Gwinner, 257–268. Springer-Verlag, Berlin.

Hagan, J. M., III. 1993. Decline of the Rufous-sided Towhee in the eastern United States. *Auk* 110:863–874.

Hagan, J. M., III, and D. W. Johnston, eds. 1992. *Ecology and conservation of Neotropical migrant landbirds.* Smithsonian Institution Press, Washington, D.C.

Hagan, J. M., III, T. L. Lloyd-Evans, and J. L. Atwood. 1991. The relationship between latitude and the timing of spring migration of North American landbirds. *Ornis Scandinavica* 22:129–136.

Hagan, J. M., III, T. L. Lloyd-Evans, J. L. Atwood, and D. S. Wood. 1992. Long-term changes in migratory landbirds in the northeastern United States. In *Ecology and conservation of Neotropical migrant landbirds,* ed. J. M. Hagan III and D. W. Johnston, 115–130. Smithsonian Institution Press, Washington, D.C.

Haila, Y. 1986. North European land birds in forest fragments: Evidence for area effects? In *Wildlife 2000: Habitat relationships of terrestrial vertebrates,* ed. J. Verner, M. Morrison, and C. J. Ralph, 315–319. University of Wisconsin Press—Madison.

Hall, G. A. 1984a. A long-term population study in an Appalachian spruce forest. *Wilson Bulletin* 96:228–240.

———. 1984b. Population decline of Neotropical migrants in an Appalachian forest. *American Birds* 38:14–18.

Hamel, P. B. 1992. Cerulean Warbler (*Dendroica cerulea*). In *Migratory nongame birds of management concern in the northeast,* ed. K. J. Schneider and D. M. Pence, 385–400. U.S. Fish and Wildlife Service, Region 5, Newton Corner, Mass.

Hamilton, T. H. 1958. Adaptive variation in the genus *Vireo. Wilson Bulletin* 70:307–346.

———. 1962. Species relationships and adaptations for sympatry in the avian genus *Vireo. Condor* 64:40–68.

Hamilton, W. J., III. 1959. Aggressive behavior in migrant Pectoral Sandpipers. *Condor* 61:161–179.

Hann, H. W. 1937. Life history of the Ovenbird in southern Michigan. *Wilson Bulletin* 49:145–237.

Harrington, B. A., P. de T. Z. Antas, and F. Silva. 1986. Northward shorebird migration on the Atlantic coast of southern Brazil. *Vida Silvestre Neotropical* 1:45–54.

Haverschmidt, F. 1968. *Birds of Surinam*. Oliver and Boyd, Edinburgh.

Hayes, F. E., P. A. Scharf, and R. S. Ridgely. 1994. Austral bird migrants in Paraguay. *Condor* 96:83–97.

Hedrick, A. V., and E. J. Temeles. 1989. The evolution of sexual dimorphism in animals: Hypotheses and tests. *Trends in Ecology and Evolution* 4(5):136–138.

Helms, C. W., and W. H. Drury. 1960. Winter and migratory weight and fat field studies on some North American buntings. *Bird-Banding* 31:1–40.

Henny, C. J., and L. J. Blus. 1986. Radiotelemetry locates wintering grounds of DDE-contaminated Black-crowned Night-Herons. *Wildlife Society Bulletin* 14:236–241.

Henny, C. J., and G. B. Herron. 1989. DDE selenium mercury and White-faced Ibis reproduction at Carson Lake, Nevada, USA. *Journal of Wildlife Management* 53:1032–1045.

Henny, C. J., F. P. Ward, K. E. Riddle, and R. M. Prouty. 1982. Migratory Peregrine Falcons, *Falco peregrinus,* accumulate pesticides in Latin America during winter. *Canadian Field-Naturalist* 96:333–338.

Henny, C. J., L. J. Blus, A. J. Krynitsky, and C. M. Bunck. 1984. Current impact of DDE on Black-crowned Night-Herons in the intermountain west. *Journal of Wildlife Management* 48:1–13.

Henny, C. J., L. J. Blus, and C. S. Hulse. 1985. Trends and effects of organochlorine residues on Oregon and Nevada wading birds, 1979–1983. *Colonial Waterbirds* 8:117–128.

Hensley, M. M., and J. B. Cope. 1951. Further data on removal and repopulation of the breeding birds in a spruce-fir forest community. *Auk* 68:483–493.

Herrera, C. M. 1978. Ecological correlates of residence and non-residence in a Mediterranean passerine bird community. *Journal of Animal Ecology* 47:871–890.

Hespenheide, H. A. 1975. Selective predation by two swifts and a swallow in Central America. *Ibis* 117:82–99.

———. 1980. Bird community structure in two Panama forests: Residents, migrants, and seasonality during the non-breeding season. In *Migrant birds in the Neotropics: Ecology, behavior, distribution, and conservation,* ed. A. Keast and E. S. Morton, 227–237. Smithsonian Institution Press, Washington, D.C.

Hilty, S. L. 1980. Relative abundance of North American temperate zone breeding migrants in western Colombia and their impact at fruiting trees. In *Migrant birds in the Neotropics: Ecology, behavior, distribution, and conservation,* ed. A. Keast and E. S. Morton, 265–271. Smithsonian Institution Press, Washington, D.C.

Hilty, S. L., and W. L. Brown. 1986. *A guide to the birds of Colombia*. Princeton University Press, Princeton, N.J.

Holdridge, L. R. 1987. *Life zone ecology*. Rev. ed. Tropical Science Center, San Jose, Costa Rica.

Holmes, R. T., and T. W. Sherry. 1988. Assessing population trends of New Hampshire forest birds: Local vs. regional patterns. *Auk* 105:756–768.

———. 1989. Ecological studies of migrant warblers in Jamaica—a progress report. *Gosse Bird Club Broadsheet* 53:7–10.

————. 1992. Site fidelity of migratory warblers in temperate breeding and Neotropical wintering areas: Implications for population dynamics, habitat selection, and conservation. In *Ecology and conservation of Neotropical migrant landbirds,* ed. J. M. Hagan III and D. W. Johnston, 563–575. Smithsonian Institution Press, Washington, D.C.

Holmes, R. T., J. C. Schultz, and P. Nothnagle. 1979. Bird predation on forest insects: An exclosure experiment. *Science* 206:462–463.

Holmes, R. T., T. W. Sherry, and F. W. Sturges. 1986. Bird community dynamics in a temperate deciduous forest: Long-term trends at Hubbard Brook. *Ecological Monographs* 56:201–220.

Holmes, R. T., T. W. Sherry, and L. Reitsma. 1989. Population structure, territoriality, and overwinter survival of two migrant warbler species in Jamaica. *Condor* 91: 545–561.

Homma, A., H. G. Schatzmayr, L. A. M. Frias, and J. A. Mesquita. 1975. Viral pollution evaluation of the Guanabara Bay, Brazil. *Revista do Instituto de Medicina Tropical, São Paulo* 17:140–145.

Howe, H. F., and D. De Steven. 1979. Fruit production, migrant bird visitation, and seed dispersal of *Guarea glabra* in Panama. *Oecologia* 39:185–196.

Howe, M. A., P. H. Geissler, and B. A. Harrington. 1989. Population trends of North American shorebirds based on the international shorebird survey. *Biological Conservation* 49:185–199.

Howell, T. R. 1971. An ecological study of the birds of the lowland pine savanna and adjacent rain forest in northeastern Nicaragua. *Living Bird* 10:185–242.

Hubbard, J. P. 1973. Avian evolution in the aridlands of North America. *Living Bird* 12:155–196.

Hudson, W. H. 1920. *Birds of La Plata.* 2 vols. J. M. Dent, London.

Hunsaker, D., II. 1972. National Parks in Colombia. *Oryx* 11:441–448.

Hunt, J. L. 1976. Sedimentation of Loiza reservoir, Puerto Rico. *SCS* (Soil Conservation Service, U.S. Department of Agriculture) 153:19.

————. 1977. Sedimentation of Cidra reservoir, Puerto Rico. *SCS* (Soil Conservation Service, U.S. Department of Agriculture) 154:14.

Hussell, D. J. T., M. H. Mather, and P. H. Sinclair. 1992. Trends in numbers of tropical and temperate wintering migrant landbirds in migration at Long Point, Ontario, 1961–1988. In *Ecology and conservation of Neotropical migrant landbirds,* ed. J. M. Hagan III and D. W. Johnston, 101–114. Smithsonian Institution Press, Washington, D.C.

Hutto, R. L. 1977. The ecology of migratory western wood warblers and the winter habitat distribution of small migratory land birds in western Mexico. Ph.D. dissertation, University of California, Los Angeles.

————. 1980. Winter habitat distribution of migratory land birds in western Mexico, with special reference to small, foliage gleaning insectivores. In *Migrant birds in the Neotropics: Ecology, behavior, distribution, and conservation,* ed. A. Keast and E. S. Morton, 181–203. Smithsonian Institution Press, Washington, D.C.

————. 1981. Seasonal variation in the foraging behavior of some migratory western wood warblers. *Auk* 98:765–777.

————. 1985a. Habitat selection by nonbreeding, migratory land birds. In *Habitat selection in birds,* ed. M. Cody, 445–476. Academic Press, New York.

————. 1985b. Seasonal changes in the habitat distribution of transient insectivorous birds in southeastern Arizona: Competition mediated? *Auk* 102:120–132.

————. 1986. Migratory landbirds in western Mexico: A vanishing habitat. *Western Wildlands* 1986:12–16.

————. 1988a. Is tropical deforestation responsible for the reported declines in Neotropical migrant populations? *American Birds* 42:375–379.

————. 1988b. Foraging behavior patterns suggest a possible cost associated with participation in mixed-species bird flocks. *Oikos* 51:79–83.

————. 1989. The effect of habitat alteration on migratory land birds in a west Mexican tropical forest: A conservation perspective. *Conservation Biology* 3:138–148.

————. 1992. Habitat distributions of migratory landbird species in western Mexico. In *Ecology and conservation of Neotropical migrant landbirds,* ed. J. M. Hagan III and D. W. Johnston, 221–239. Smithsonian Institution Press, Washington, D.C.

————. 1994. The composition and social organization of mixed-species flocks in a tropical deciduous forest in western Mexico. *Condor* 96:105–118.

ICAITI. 1977. *An environmental and economic study of the consequences of pesticide use in Central American cotton production: Final report.* Instituto Centro Americano de Investigación y Tecnología Industrial, Guatemala.

James, F. C., D. A. Wiedenfeld, and C. E. McCulloch. 1992. Trends in breeding populations of warblers: Declines in the southern highlands and increases in the lowlands. In *Ecology and conservation of Neotropical migrant landbirds,* ed. J. M. Hagan III and D. W. Johnston, 43–56. Smithsonian Institution Press, Washington, D.C.

Janzen, D. H. 1980. Heterogeneity of potential food abundance for tropical small land birds. In *Migrant birds in the Neotropics: Ecology, behavior, distribution, and conservation,* ed. A. Keast and E. S. Morton, 545–552. Smithsonian Institution Press, Washington, D.C.

Jehl, J. R., Jr. 1990. Aspects of the molt migration. In *Bird migration: Physiology and ecophysiology,* ed. E. Gwinner, 102–113. Springer-Verlag, Berlin.

Jenni, L., and S. Jenni-Eiermann. 1992. Metabolic patterns of feeding, overnight fasted, and flying night migrants during autumn migration. *Ornis Scandinavica* 23:251–259.

Johnson, A. W., and J. D. Goodall. 1965. *The birds of Chile and adjacent regions of Argentina, Bolivia, and Peru.* Vol. 1. Platt Establecimientos Gráficos, Buenos Aires.

————. 1967. *The birds of Chile and adjacent regions of Argentina, Bolivia, and Peru.* Vol. 2. Platt Establecimientos Gráficos, Buenos Aires.

Johnson, D. H., G. L. Krapu, K. H. Reinecke, and D. G. Jorde. 1985. An evaluation of condition indices for birds. *Journal of Wildlife Management* 49:569–575.

Johnson, J. A., and F. R. Ziegler. 1978. A Violet-crowned Hummingbird in California. *Western Birds* 9:91–92.

Johnson, N. K., and R. M. Zink. 1985. Genetic evidence for relationships among the Red-eyed, Yellow-green, and Chivi vireos. *Wilson Bulletin* 97:421–435.

Johnson, S. R., and W. J. Richardson. 1982. Waterbird migration near the Yukon and Alaskan coast of the Beaufort Sea. 2. Moult migration of seaducks in summer. *Arctic* 35:291–301.

Johnson, T. B. 1980. Resident and North American migrant bird interactions in the Santa Marta highlands, Northern Colombia. In *Migrant birds in the Neotropics: Ecology, behavior, distribution, and conservation,* ed. A. Keast and E. S. Morton, 239–247. Smithsonian Institution Press, Washington, D.C.

Johnston, D. W., and D. L. Winings. 1987. Natural history of Plummers Island, Maryland. 27. The declines of forest birds on Plummers Island, Maryland, and vicinity. *Proceedings of the Biological Society of Washington* 100:762–768.

Jones, E. T. 1986. The passerine decline. *North American Bird Bander* 11:74–75.

Jones, P. J. 1985. The migration strategies of Palaearctic passerines in West Africa. In *Migratory birds: Problems and prospects in Africa,* ed. A. MacDonald and P. Goriup, 9–21. Report of the Fourteenth Conference European Continental Shelf Section, 1983. International Council for Bird Preservation, Cambridge.

Jukofsky, D. 1993. Mystical messenger. *Nature Conservancy* 43(6):24–29.

Julin, A. M., and H. O. Sanders. 1977. Toxicity and accumulation of the insecticide imidan in fresh water invertebrates and fishes. *Transactions of the American Fisheries Society* 106:386–392.

Junge, G. C. A., and K. H. Voous. 1955. The distribution and relationship of *Sterna eurygnatha* Saunders. *Ardea* 43:226–247.

Kale, H. W., II. 1967. Aggressive behavior by a migrating Cape May Warbler. *Auk* 84:120–121.

———. 1971a. Structure of avian communities in selected Panama and Illinois habitats. *Ecological Monographs* 41:207–230.

———. 1971b. Wintering Kentucky Warblers. *Bird-Banding* 42:299.

———. 1976. On the relative abundance of migrants from the North Temperate Zone in tropical habitats. *Wilson Bulletin* 88:433–458.

———. 1978. Man and wildlife in the Tropics: Past, present, and future. In *Wildlife and people,* John S. Wright Forest Conference Proceedings, 120–139. Department of Forestry and Natural Resources, Cooperative Extension Service, Purdue University, Lafayette, Ind.

———. 1981. Surveying birds in the Tropics. *Studies in Avian Biology* 6:548–553.

Karr, J. R., J. D. Nichols, M. K. Klimkiewicz, and J. D. Brawn. 1990. Survival rates of birds in tropical and temperate forests: Will the dogma survive? *American Naturalist* 136:277–291.

Kaufmann, J. H. 1983. On the definitions and functions of dominance and territoriality. *Biological Review* 58:1–20.

Keast, A. 1980. Spatial relationships between migratory parulid warblers and their ecological counterparts in the Neotropics. In *Migrant birds in the Neotropics: Ecology, behavior, distribution, and conservation,* ed. A. Keast and E. S. Morton, 109–130. Smithsonian Institution Press, Washington, D.C.

Keast, A., and E. S. Morton, eds. 1980. *Migrant birds in the Neotropics: Ecology, behavior, distribution, and conservation.* Smithsonian Institution Press, Washington, D.C.

Kendeigh, S. C. 1982. *Bird populations in east-central Illinois: Fluctuations, variations, and development over a half century.* Illinois Biological Monographs, no. 52.

Kerlinger, P., and F. R. Moore. 1989. Atmospheric structure and avian migration. *Current Ornithology* 6:109–142.

Ketterson, E. D., and V. Nolan, Jr. 1976. Geographic variation and its climatic correlates in the sex ratio of eastern-wintering Dark-eyed Juncos (*Junco hyemalis hyemalis*). *Ecology* 57:679–693.

———. 1983. The evolution of differential bird migration. *Current Ornithology* 1: 357–402.

Kilham, L. 1978. Sexual similarity of Red-headed Woodpeckers and possible explanations based on fall territorial behavior. *Wilson Bulletin* 90:285.

King, B., and E. C. Dickinson. 1975. *A field guide to the birds of South-east Asia.* Collins, London.

King, J. R. 1961. The bioenergetics of vernal premigratory fat deposition in the White-crowned Sparrow. *Condor* 63:128–142.

———. 1972. Adaptive periodic fat storage by birds. *Proceedings of the International Ornithological Congress* 15:200–217.

King, J. R., and D. S. Farner. 1963. The relationship of fat deposition to Zugunruhe and migration. *Condor* 65:200–223.

Klopfer, P. H., and R. H. MacArthur. 1961. On the causes of tropical species diversity: Niche overlap. *American Naturalist* 95:223–226.

Klopfer, P. H., D. I. Rubenstein, R. S. Eigely, and R. J. Barnett. 1974. Migration and species diversity in the Tropics. *Proceedings of the National Academy of Sciences* (USA) 71:339–340.

Kodric-Brown, A., and J. H. Brown. 1978. Influence of economics, interspecific competition, and sexual dimorphism on territoriality of migrant Rufous Hummingbirds. *Ecology* 59:285–296.

Koolhaus, M. H. 1977. The universal soil loss equation, Montevideo, Uruguay [in Spanish]. *Boletín de la Universidad de la República, Facultad de Agronomía* 130:37.

Krebs, J. R. 1971. Territory and breeding density in the Great Tit, *Parus major* L. *Ecology* 52:2–22.

Kricher, J. C., and W. E. Davis, Jr. 1992. Patterns of avian species richness in disturbed and undisturbed habitats in Belize. In *Ecology and conservation of Neotropical migrant landbirds,* ed. J. M. Hagan III and D. W. Johnston, 240–246. Smithsonian Institution Press, Washington, D.C.

Lack, D. 1944. Ecological aspects of species formation in passerine birds. *Ibis* 86: 260–286.

———. 1954. *The natural regulation of animal numbers.* Oxford University Press, London.

————. 1956. A review of the genera and nesting habits of swifts. *Auk* 73:1–32.

————. 1960. The influence of weather on passerine migration: A review. *Auk* 77: 171–209.

————. 1968a. Bird migration and natural selection. *Oikos* 19:1–9.

————. 1968b. *Ecological adaptations for breeding in birds.* Methuen, London.

Lack, D., and P. C. Lack. 1972. Wintering warblers in Jamaica. *Living Bird* 11:129–153.

Lack, P. C. 1986. Ecological correlates of migrants and residents in a tropical African savanna. *Ardea* 74:111–119.

Lacombe, D., and W. Moneiro. 1975. Balanidae as pollution indicators in the Bay of Guanabara, Brazil. *Revista Brasiliera Biológica* 34:633–644.

Lanyon, W. E. 1978. Revision of the *Myiarchus* flycatchers of South America. *Bulletin of the American Museum of Natural History* 161:427–628.

————. 1982. Evidence for wintering and resident populations of Swainson's Flycatcher (*Myiarchus swainsoni*) in northern Surinam. *Auk* 99:581–582.

LaPerriere, A. J., and A. O. Haugen. 1972. Some factors influencing calling activity of wild Mourning Doves. *Journal of Wildlife Management* 36:1193–1199.

Lara Madrid, C., and E. N. Razetti. 1971. Sea water pollution in Venezuela. *FAO Fish Report* (Food and Agricultural Organization, United Nations) 99:117.

Larkin, R. R., D. R. Griffin, J. R. Torre-Bueno, and J. M. Teal. 1979. Radar observations of bird migration over the western North Atlantic Ocean. *Behavioral Ecology and Sociobiology* 4:225–264.

Larson, S. 1957. The suborder Charadrii in arctic and boreal areas during the Tertiary and Pleistocene. *Acta Vertebrata* 1:1–81.

Laviada, I. 1976. Deforestation and soil erosion in Mexico, Central America [in Spanish]. *Mexican Forestry* 50:27–28.

Leck, C. F. 1972a. The impact of some North American migrants at fruiting trees in Panama. *Auk* 89:842–850.

————. 1972b. Observations of birds at *Cecropia* trees in Puerto Rico. *Wilson Bulletin* 84:498–500.

————. 1972c. Seasonal changes in feeding pressures of fruit- and nectar-eating birds in Panama. *Condor* 74:54–60.

Leck, C. F., B. G. Murray, Jr., and J. Swineboard. 1981. Changes in breeding bird populations at Hutcheson Memorial Forest since 1958. *William L. Hutcheson Memorial Forest Bulletin* 6:8–14.

Lederer, R. J. 1977. Winter feeding territories in the Townsend's Solitaire. *Bird-Banding* 48:11–18.

LeGrand, H. E., and K. J. Schneider. 1992. Bachman's Sparrow (*Aimophila aestivalis*). In *Migratory nongame birds of management concern in the northeast,* ed. K. J. Schneider and D. M. Pence, 299–313. U.S. Fish and Wildlife Service, Region 5, Newton Corner, Mass.

Leimgruber, P., W. J. McShea, and J. H. Rappole. 1994. Predation on artificial nests in large forest blocks. *Journal of Wildlife Management* 58:254–260.

Leisler, B. 1990. Selection and use of habitat of wintering migrants. In *Bird migration: Physiology and ecophysiology,* ed. E. Gwinner, 156–174. Springer-Verlag, Berlin.

Lekagul, B. 1968. *Bird guide of Thailand.* Association for Conservation of Wildlife, Bangkok.

Lenon, H., L. V. Curry, A. Miller, and D. Patulski. 1972. Insecticide residues in water and sediment from cisterns on the USA and British Virgin Islands, 1970. *Pesticides Monitoring Journal* 6:188–193.

Leopold, N. F., Jr. 1963. *Checklist of birds of Puerto Rico and the Virgin Islands.* University of Puerto Rico, Agricultural Experiment Station Bulletin no. 168.

Levey, D. J., and F. G. Stiles. 1992. Evolutionary precursors of long-distance migration: Resource availability and movement patterns in Neotropical landbirds. *American Naturalist* 140:467–491.

Levins, R. 1968. *Evolution in changing environments.* Princeton University Press, Princeton, N.J.

Lincer, J. L., and J. A. Sherburne. 1974. Organochlorines in kestrel prey: A north–south dichotomy. *Journal of Wildlife Management* 38:427–434.

Lincoln, F. C. 1936. Returns of banded birds. Third paper. Some recoveries of water birds from Latin America. *Bird-Banding* 7:139–148.

Lindstrom, A. 1990. The role of predation risk in stopover habitat selection in migrating Bramblings, *Fringilla montifringilla. Behavioral Ecology* 1:102–106.

Line, L. 1993. Silence of the songbirds. *National Geographic* 183(6):68–90.

Litwin, T. S. 1986. Factors affecting avian diversity in a northeastern woodlot. Ph.D. dissertation, Cornell University, Ithaca, N.Y.

Litwin, T. S., and C. R. Smith. 1992. Factors influencing the decline of Neotropical migrants in a northeastern forest fragment: Isolation, fragmentation, or mosaic effects? In *Ecology and conservation of Neotropical migrant landbirds,* ed. J. M. Hagan III and D. W. Johnston, 483–496. Smithsonian Institution Press, Washington, D.C.

Loetscher, F. W., Jr. 1941. Ornithology of the Mexican state of Veracruz with an annotated list of the birds. Ph.D. dissertation, Cornell University Press, Ithaca, N.Y.

Loftin, H. 1977. Returns and recoveries of banded North American birds in Panama and the Tropics. *Bird-Banding* 48:253–258.

Loftin, H., G. I. Child, and S. Bongiorno. 1967. Returns in 1965–1966 of North American migrant birds banded in Panama. *Bird-Banding* 38:151–152.

Loiselle, B. A. 1987. Migrant abundance in a Costa Rican lowland forest canopy. *Journal of Tropical Ecology* 3:163–168.

Loiselle, B. A., and J. G. Blake. 1991. Resource abundance and temporal variation in fruit-eating birds along a wet forest elevational gradient in Costa Rica. *Ecology* 72:180–193.

Lopez Ornat, A., and R. Greenberg. 1990. Sexual segregation by habitat in migratory warblers in Quintana Roo, Mexico. *Auk* 107:539–543.

Lopez Saucedo, M. 1975. Only the sixth part of Mexico without erosion. *Mexican Forestry* 49:27–28.

Lovejoy, T. E., III. 1981. A world less green. *Defenders* 56:2–5.

Lowery, G. H., Jr. 1945. Trans-Gulf migration of birds and the coastal hiatus. *Wilson Bulletin* 57:97–121.

———. 1946. Evidence of trans-Gulf migration. *Auk* 63:175–211.

———. 1951. A quantitative study of the nocturnal migration of birds. *University of Kansas Publications of the Museum of Natural History* 3:361–472.

———. 1955. *Louisiana birds.* Louisiana State University Press, Baton Rouge.

Lum, A. L. 1978. Shorebird fauna changes of a small tropical estuary following habitat alteration: Biological and political impacts of environmental restoration. *Environmental Management* 2(5):423–430.

Lynch, J. F. 1989. Distribution of overwintering Nearctic migrants in the Yucatán Peninsula. 1. General patterns of occurrence. *Condor* 91:515–544.

———. 1992. Distribution of overwintering Nearctic migrants in the Yucatán Peninsula. 2. Use of native and human modified vegetation. In *Ecology and conservation of Neotropical migrant landbirds,* ed. J. M. Hagan III and D. W. Johnston, 178–196. Smithsonian Institution Press, Washington, D.C.

Lynch, J. F., and R. F. Whitcomb. 1978. Effects of the insularization of the eastern deciduous forest on avifaunal diversity and turnover. In *Classification, inventory, and analysis of fish and wildlife habitat,* ed. A. Marmelstein, 461–489. U.S. Fish and Wildlife Service, Washington, D.C.

Lynch, J. F., E. S. Morton, and M. E. Van der Voort. 1985. Habitat segregation between the sexes of wintering Hooded Warblers (*Wilsonia citrina*). *Auk* 102: 714–721.

Mabey, S. E., and E. S. Morton. 1992. Demography and territorial behavior of wintering Kentucky Warblers in Panama. In *Ecology and conservation of Neotropical migrant landbirds,* ed. J. M. Hagan III and D. W. Johnston, 329–336. Smithsonian Institution Press, Washington, D.C.

MacArthur, R. H. 1958. Population ecology of some warblers of northeastern coniferous forest. *Ecology* 39:599–619.

———. 1972. *Geographical ecology: Patterns in the distribution of species.* Harper and Row, New York.

MacArthur, R. H., and E. O. Wilson. 1967. *The theory of island biogeography.* Princeton University Press, Princeton, N.J.

MacArthur, R. H., H. F. Recher, and M. L. Cody. 1966. On the relation between habitat selection and species diversity. *American Naturalist* 100:319–332.

MacKenzie-Grieve, R. C., and J. B. Tatum. 1974. Costa's Hummingbird: A new bird for Canada. *Canadian Field* 88:91–92.

Marantz, C. A., and J. V. Remsen, Jr. 1991. Seasonal distribution of the Slaty Elaenia, a little-known austral migrant of South America. *Journal of Field Ornithology* 62:162–172.

Marke, M. 1906. Einfluss von Wind und Wetter auf den Vogelzug. *Ornithologie Jahrbuch* 17 (Suppl.): 81–136, 161–199.

Margolis, E., A. Vieira de Mello Netto, I. de A. Albuquerque, M. Montenegro, Jr., and

G. De A. B. Campello. 1975. Survey of capacity of land use and conservation planning of the Tambe experiment, Brazil [in Portuguese]. *Boletín Técnico del Instituto de Pesquisa Agronomía* 73:36.

Marion, W. R. 1974. Ecology of the Plain Chachalaca in the lower Rio Grande Valley of Texas. Ph.D. dissertation, Texas A&M University, College Station.

Marshall, J. R. 1988. Birds lost from a giant sequoia forest during fifty years. *Condor* 90:359–372.

Martin, T. E. 1987. Food as a limit on breeding birds: A life history perspective. *Annual Review of Ecology and Systematics* 18:453–487.

Martinez, J. 1973. Erosion threatens 1,444,000 more in Alto Papaloapan [in Spanish]. *Mexican Forestry* 47:26.

Mayr, E. 1946. History of the North American bird fauna. *Wilson Bulletin* 58:2–41.

———. 1964. Neotropical region. In *New dictionary of birds,* ed. A. L. Thomson, 516–518. McGraw-Hill, New York.

Mayr, E., and W. Meise. 1930. Theories on the history of migrants [in German]. *Vogelzug* 1:149–172.

Mayr, E., and L. L. Short, Jr. 1970. *Species taxa of North American birds.* Publications of the Nuttall Ornithological Club, no. 9.

McCabe, R. E., and T. R. McCabe. 1984. Of slings and arrows: An historical retrospection. In *White-tailed deer: Ecology and management,* ed. L. K. Halls, 19–72. Stackpole Books, Harrisburg, Pa.

McClintock, C. P., T. C. Williams, and J. M. Teal. 1978. Autumnal bird migration observed from ships in the western North Atlantic Ocean. *Bird-Banding* 49:262–277.

McClure, H. E. 1974. *Migration and survival of the birds of Asia.* U.S. Army Medical Component, SEATO Medical Project, Bangkok.

McDiarmid, R. W., R. E. Ricklefs, and M. S. Foster. 1977. Dispersal of *Stommadenia donnell-smithii* (Apocynaceae) by birds. *Biotropica* 9:9–25.

McDonald, M. V., E. S. Morton, and S. Mabey. n.d. Behavioral ecology of the Kentucky Warbler (*Oporornis formosus*) in northern Virginia oak forest. Conservation and Research Center, Front Royal, Va. Manuscript.

McNeil, R., and M. Carrera de Itriago. 1968. Fat deposition in the Scissor-tailed Flycatcher (*Muscivora t. tyrannus*) and the Small-billed Elaenia (*Elaenia parvirostris*) during the austral migratory period in northern Venezuela. *Canadian Journal of Zoology* 46:123–128.

McShea, W. J., and J. H. Rappole. 1992. White-tailed deer as keystone species within forested habitats of Virginia. *Virginia Journal of Science* 43:177–186.

McShea, W. J., J. H. Rappole, and G. Burford. n.d. Variable song rates for three species of passerines and significance for their conservation. Conservation and Research Center, Front Royal, Va. Manuscript.

Medel y Alvarado, L. 1963. *History of San Andrés Tuxtla: 1532–1950* [in Spanish]. Editorial Citlaltepetl, Tacubaya, Mexico City.

Mengel, R. M. 1964. The probable history of species formation in some northern wood warblers. *Living Bird* 3:9–44.

————. 1970. The North American central plains as an isolating agent in bird speciation. In *Pleistocene recent environments of the central Great Plains,* ed. W. Dort, Jr., and J. K. Jones, 279–340. Department of Geology, University of Kansas, Special Publication no. 3.

Meyer de Schauensee, R. 1966. *The species of birds of South America and their distribution.* Livingston, Wynnewood, Pa.

Meyer de Schauensee, R., and W. H. Phelps, Jr. 1978. *A guide to the birds of Venezuela.* Princeton University Press, Princeton, N.J.

Miller, A. H. 1963. *Seasonal activity and ecology of the avifauna of an American equatorial cloud forest.* University of California Publications in Zoology, vol. 66. Berkeley.

Mlecko, B. E. 1968. Notes on the birds of Costa Rica with special emphasis on flocking. *Proceedings of the Iowa Academy of Sciences* 75:457–462.

Monkkonen, M., P. Helle, and D. Welsh. 1992. Perspectives on Palaearctic and Nearctic bird migration: Comparisons and overview of life-history and ecology of migrant passerines. *Ibis* 134 (Suppl. 1): 7–13.

Moore, F. R., and T. R. Simons. 1992. Habitat suitability and stopover ecology of Neotropical landbird migrants. In *Ecology and conservation of Neotropical migrant landbirds,* ed. J. M. Hagan III and D. W. Johnston, 345–355. Smithsonian Institution Press, Washington, D.C.

Moore, T. S. 1974. First record of the Smooth-billed Ani in Georgia. *Oriole* 40:1–2.

Morain, S. A. 1984. *Systematic and regional biogeography.* Van Nostrand Reinhold, New York.

Moreau, R. E. 1952. The place of Africa in the Palaearctic migration system. *Journal of Animal Ecology* 21:250–271.

————. 1966. The mutability of the African avifaunal scene. *Ostrich* (Suppl. 6): 453–459.

————. 1972. *The Palaearctic–African bird migration system.* Academic Press, New York.

Morel, G., and F. Bourlière. 1962. Ecological relations of the sedentary and migratory avifauna in a Sahel savannah of lower Senegal [in French]. *Terre et Vie* 4:371–393.

Morrison, R. I. G. 1984. Migration systems of some New World shorebirds. In *Behavior of marine animals,* ed. J. Burger, and B. L. Olla, 125–202. Plenum Press, New York.

Morrison, R. I. G., and J. P. Myers. 1989. Shorebird flyways in the New World. In *Flyways and reserve networks for waterbirds,* ed. H. Boyd and J.-Y. Pirot, 85–96. IWRB, Slimbridge, U.K.

Morse, D. H. 1970. Ecological aspects of some mixed-species foraging flocks of birds. *Ecological Monographs* 40:119–168.

————. 1971. The insectivorous bird as an adaptive strategy. *Annual Review of Ecology and Systematics* 2:177–200.

————. 1989. *American warblers.* Harvard University Press, Cambridge.

Morton, E. S. 1971. Food and migration habits of the Eastern Kingbird in Panama. *Auk* 88:925–926.

———. 1972. North American birds in the Tropics. *Atlantic Naturalist* 27:164–168.

———. 1973. On the evolutionary advantages and disadvantages of fruit eating in tropical birds. *American Naturalist* 107:8–22.

———. 1976. The adaptive significance of dull-coloration in Yellow Warblers. *Condor* 78:423.

———. 1977. Intratropical migration in the Yellow-green Vireo and Piratic Flycatcher. *Auk* 94:97–106.

———. 1979. Effective pollination of *Erythrina fusca* by the Orchard Oriole (*Icterus spurius*): Coevolved behavioral manipulation? *Annals of the Missouri Botanical Gardens* 66:482–489.

———. 1980. Adaptations to seasonal changes by migrant land birds in the Panama Canal Zone. In *Migrant birds in the Neotropics: Ecology, behavior, distribution, and conservation,* ed. A. Keast and E. S. Morton, 437–453. Smithsonian Institution Press, Washington, D.C.

———. 1990. Habitat segregation by sex in the Hooded Warbler: Experiments on proximate causation and discussion of its evolution. *American Naturalist* 135: 319–333.

———. 1992. What do we know about the future of migrant landbirds? In *Ecology and conservation of Neotropical migrant landbirds,* ed. J. M. Hagan III and D. W. Johnston, 579–589. Smithsonian Institution Press, Washington, D.C.

Morton, E. S., and R. Greenberg. 1989. The outlook for migratory songbirds: "Future shock" for birders. *American Birds* 43:178–183.

Morton, E. S., M. Van der Voort, and R. Greenberg. 1993. How a warbler chooses its habitat: Field support for laboratory experiments. *Animal Behaviour* 46:47–53.

Morton, E. S., J. F. Lynch, K. Young, and P. Melhop. 1987. Do male Hooded Warblers exclude females from non-breeding territories in tropical forest? *Auk* 104: 133–135.

Moynihan, M. 1962. *The organization and probable evolution of some mixed species flocks of Neotropical birds.* Smithsonian Miscellaneous Collections, vol. 143. Washington, D.C.

Munn, C. A. 1985. Permanent canopy and understory flocks in Amazonia: Species composition and population density. *Ornithological Monographs* 36:683–712.

Munn, C. A., and J. W. Terborgh. 1979. Multispecies territoriality in Neotropical foraging flocks. *Condor* 81:338–347.

Murray, B. G., Jr. 1965. On the autumn migration of the Blackpoll Warbler. *Wilson Bulletin* 77:122–133.

———. 1966. Migration of age and sex classes of passerines on the Atlantic coast in autumn. *Auk* 83:352–360.

———. 1976. The return to the mainland of some nocturnal passerine migrants over the sea. *Bird-Banding* 47:345–358.

———. 1989. A critical review of the Transoceanic migration of the Blackpoll Warbler. *Auk* 106:8–17.

Myers, J. P. 1980. The pampas shorebird community: Interactions between breeding and nonbreeding members. In *Migrant birds in the Neotropics: Ecology, behavior, distribution, and conservation,* ed. A. Keast and E. S. Morton, 37–49. Smithsonian Institution Press, Washington, D.C.

————. 1981. A test of three hypotheses for latitudinal segregation of the sexes in wintering birds. *Canadian Journal of Zoology* 59:1527–1534.

Myers, J. P., P. G. Conners, and F. A. Pitelka. 1979. Territoriality in non-breeding shorebirds: Shorebirds in marine environments. *Studies in Avian Biology* 2:231–246.

Myers, J. P., G. Castro, B. Harrington, M. Howe, J. Maron, E. Ortiz, M. Sallaberry, C. T. Schick, and E. Tabilo. 1984. The Pan American shorebird program: A progress report. *Wader Study Group Bulletin* 42:26–31.

Myers, J. P., R. I. G. Morrison, P. A. Antas, B. A. Harrington, T. E. Lovejoy, M. Sallaberry, S. E. Senner, and A. Tarak. 1987. Conservation strategy for migratory species. *American Scientist* 75:19–26.

Myers, N. 1980a. The conversion of tropical forests. *Environment* 22:6–13.

————. 1980b. *Conversion of tropical moist forests.* National Research Council, Committee on Research Priorities in Tropical Biology. National Academy of Sciences, Washington, D.C.

————. 1980c. The present status and future prospects of tropical moist forests. *Environmental Conservation* 7:101–114.

Nations, J. D., and R. B. Nigh. 1978. Cattle, cash food, and forest: The destruction of the American Tropics and the Lacandones Maya alternative. *Culture and Agriculture* 6:1–5.

Nice, M. M. 1964. *Studies of the life history of the Song Sparrow.* Vol. 2, *Behavior.* Dover Publications, New York.

Nickell, W. P. 1968. Return of northern migrants to tropical winter quarters and banded birds recovered in the United States. *Bird-Banding* 39:107–116.

Nisbet, I. C. T. 1970. Autumn migration of the Blackpoll Warbler: Evidence for long flight provided by regional survey. *Bird-Banding* 41:207–240.

Nisbet, I. C. T., and W. H. Drury. 1968. Short-term effects of weather on bird migration: A field study using multivariate statistics. *Animal Behaviour* 16:496–530.

Nisbet, I. C. T., W. H. Drury, and J. Baird. 1963. Weight loss during migration. Part 1. Deposition and consumption of fat by the Blackpoll Warbler (*Dendroica striata*). *Bird-Banding* 34:107–138.

Nisbet, I. C. T., and L. Medway. 1972. Dispersion, population ecology, and migration of eastern great reed warblers *Acrocephalus orientalis* wintering in Malaysia. *Ibis* 114:451–494.

Nolan, V., Jr. 1978. The ecology and behavior of the Prairie Warbler *Dendroica discolor. Ornithological Monographs,* no. 26.

Novak, P. G. 1992. Black Tern (*Chlidonias niger*). In *Migratory nongame birds of management concern in the northeast,* ed. K. J. Schneider and D. M. Pence, 149–170. U.S. Fish and Wildlife Service, Region 5, Newton Corner, Mass.

Odum, E. P., C. E. Connell, and H. L. Stoddard. 1961. Flight energy and estimated flight ranges of some migratory birds. *Auk* 78:515–527.

Ollson, L. 1985. *An integrated study of desertification*. C. W. K. Gleerup, Malmö, Sweden.

Olrog, C. C. 1963a. Banding of birds in Argentina, 1961–1963 [in Spanish]. Part 3. *Neotropica* 9 (Suppl.): 1–8.

———. 1963b. *List and distribution of Argentine birds* [in Spanish]. Universidad Nacional Tucuman, Instituto Miguel Lillo, Opera Lilliano 9.

———. 1968. Banding of birds in Argentina, 1964–1966 [in Spanish]. Part 5. *Neotropica* 14:17–22.

———. 1969. Birds of South America. In *Biogeography and ecology in South America,* ed. E. J. Fittkau, J. Illies, H. Klinge, G. H. Schwabe, and H. Sioli, 849–878. Monographiae Biologicae, Junk, The Hague.

———. 1975. The banding of birds in Argentina, 1961–1974 [in Spanish]. *Neotropica* 21:17–19.

Oniki, Y. 1979. Is nesting success of birds low in the Tropics? *Biotropica* 11:60–69.

Orejuela, J. E., R. J. Raitt, and H. Alvarez. 1980. Differential use by North American migrants of Colombian forests. In *Migrant birds in the Neotropics: Ecology, behavior, distribution, and conservation,* ed. A. Keast and E. S. Morton, 253–264. Smithsonian Institution Press, Washington, D.C.

Orians, G. H. 1969. The number of bird species in some tropical forests. *Ecology* 50:783–801.

Osborne, D. R., and A. T. Peterson. 1984. Decline of the Upland Sandpiper (*Bartramia longicauda*) in Ohio: An endangered species. *Ohio Journal of Science* 84(1):8–10.

Palmer, R. S. 1962, 1975, 1976. *Handbook of North American Birds.* Vols. 1–3. Yale University Press, New Haven, Conn.

Parker, T. A., III, S. A. Parker, and M. A. Plenge. 1982. *An annotated checklist of Peruvian birds.* Buteo Books, Vermillion, S.Dak.

Parnell, J. F. 1969. Habitat relations of the Parulidae during spring migration. *Auk* 86:505–521.

Parrish, J. D., and T. W. Sherry. 1994. Sexual habitat segregation by American Redstarts wintering in Jamaica: Importance of resource seasonality. *Auk* 111:38–49.

Payne, R. B. 1984. Sexual selection, lek behavior, and sexual size dimorphism in birds. *Ornithological Monographs,* no. 33.

Paynter, R. A., Jr. 1955. *The ornithogeography of the Yucatán Peninsula.* Peabody Museum of Natural History, Bulletin no. 9.

Pearson, D. L. 1980. Bird migration in Amazonian Ecuador, Peru, and Bolivia. In *Migrant birds in the Neotropics: Ecology, behavior, distribution, and conservation,* ed. A. Keast and E. S. Morton, 273–283. Smithsonian Institution Press, Washington, D.C.

Pennington, T. D., and J. Sarukhan. 1968. *Tropical trees of Mexico* [in Spanish]. Instituto Nacional de Investigaciones Forestales, Mexico City.

Peralta, P. M. 1977. *Gullies and their control* [in Spanish]. Manual, Facultad de Ciencias, Universidad de Chile, no. 5.

Peters, J. L., with collaborators. 1931–1986. *Checklist of birds of the world.* 15 vols. Harvard University Press, Cambridge.

Peterson, J. M. C., and C. Fichtel. 1992. Olive-sided Flycatcher (*Contopus borealis*). In *Migratory nongame birds of management concern in the northeast,* ed. K. J. Schneider and D. M. Pence, 149–170. U.S. Fish and Wildlife Service, Region 5, Newton Corner, Mass.

Peterson, R. T. 1980. *A field guide to the birds.* Houghton Mifflin, Boston.

Peterson, R. T., and E. Chalif. 1989. *Birds of Mexico* [in Spanish]. Trans. M. A. Ramos and M. I. Castillo. World Wildlife Fund, Editorial Diana, Mexico City.

Peterson, R. T., G. Mountfort, and P. A. D. Hollom. 1967. *A field guide to the birds of Britain and Europe.* 2d ed. Houghton Mifflin, Boston.

Petit, D. R., L. J. Petit, and K. G. Smith. 1992. Habitat associations of migratory birds overwintering in Belize, Central America. In *Ecology and conservation of Neotropical migrant landbirds,* ed. J. M. Hagan III and D. W. Johnston, 247–256. Smithsonian Institution Press, Washington, D.C.

Petit, D. R., J. F. Lynch, R. L. Hutto, J. G. Blake, and R. B. Waide. 1993. Management and conservation of migratory landbirds overwintering in the Neotropics. In *Status and management of Neotropical migratory birds,* ed. D. M. Finch and P. W. Stangel, 70–92. U.S. Department of Agriculture, Forest Service, General Technical Report RM-229. Rocky Mountain Forest and Range Experiment Station, Fort Collins, Colo.

———. 1995. Habitat use and conservation during winter in the Neotropics. In *Ecology and management of Neotropical migratory birds: A synthesis and review of the critical issues,* ed. T. Martin and D. Finch. Oxford University Press, New York.

Phelps, W. H., and W. H. Phelps, Jr. 1958. *List of the birds of Venezuela with their distribution* [in Spanish]. Vol. 1, part 1, *No Passeriformes.* Boletín de la Sociedad Venezolana de Ciencias Naturales, no. 19.

———. 1963. *List of the birds of Venezuela with their distribution* [in Spanish]. Vol. 1, part 2, *Passeriformes.* Boletín de la Sociedad Venezolana de Ciencias Naturales, no. 24.

Phillips, A. R. 1951. Complexities of migration: A review. *Wilson Bulletin* 63:129–136.

Phillips, A. R., J. T. Marshall, and G. Monson. 1964. *The birds of Arizona.* University of Arizona Press, Tucson.

Post, P. W. 1978. Social and foraging behavior of warblers wintering in Puerto Rican coastal scrub. *Wilson Bulletin* 90:197–214.

Powell, G. V. N. 1980. Migrant participation in Neotropical mixed species flocks. In *Migrant birds in the Neotropics: Ecology, behavior, distribution, and conservation,* ed. A. Keast and E. S. Morton, 477–483. Smithsonian Institution Press, Washington, D.C.

———. 1985. Sociobiology and adaptive significance of interspecific foraging flocks in the Neotropics. *Ornithological Monographs* 36:713–732.

Powell, G. V. N., and R. Bjork. 1994. Implications of altitudinal migration for conservation strategies to protect tropical biodiversity: A case study at Monteverde, Costa Rica. *Bird Conservation International* 4:161–174.

Powell, G. V. N., and H. L. Jones. 1978. An observation of polygyny in the Common Yellowthroat. *Wilson Bulletin* 90:656–657.

Powell, G. V. N., and J. H. Rappole. 1986. The Hooded Warbler. In *Audubon Wildlife Report,* vol. 3, ed. R. L. Di Silvestro, 827–853. National Audubon Society, New York.

Powell, G. V. N., J. H. Rappole, and S. A. Sader. 1992. Nearctic migrant use of lowland Atlantic habitats in Costa Rica: A test of remote sensing for identification of habitat. In *Ecology and conservation of Neotropical migrant landbirds,* ed. J. M. Hagan III and D. W. Johnston, 287–298. Smithsonian Institution Press, Washington, D.C.

Power, D. M. 1971. Warbler ecology: Diversity, similarity, and seasonal differences in habitat segregation. *Ecology* 52:434–443.

Pregnolatto, W., N. S. Garrido, and M. De Toledo. 1974. Research and detection of mercury residues in Brazilian salt and fresh water fishes. *Revista del Instituto Adolfo Lutz* 34:95–100.

Pulich, W. M. 1976. *The Golden-cheeked Warbler: A bioecological study.* Texas Parks and Wildlife Department, Austin.

Pycraft, W. P. 1910. *A history of birds.* Methuen, London.

Rabenold, K. N. 1980. The Black-throated Green Warbler in Panama: A geographic and seasonal comparison of foraging. In *Migrant birds in the Neotropics: Ecology, behavior, distribution, and conservation,* ed. A. Keast and E. S. Morton, 297–307. Smithsonian Institution Press, Washington, D.C.

Rabol, J. 1987. Coexistence and competition between over-wintering Willow Warblers (*Phylloscopus trochilus*) and local warblers at Lake Naivasha, Kenya. *Ornis Scandinavica* 18:101–121.

Ralph, C. J. 1978. Disorientation and possible fate of young passerine coastal migrants. *Bird-Banding* 49:237–247.

Ramos, M. A. 1983. Seasonal movements of bird populations at a Neotropical study site in southern Veracruz, Mexico. Ph.D. dissertation, University of Minnesota, Minneapolis.

———. 1988. Eco-evolutionary aspects of bird movements in the northern Neotropical region. *Proceedings of the International Ornithological Congress* 19:251–293.

Ramos, M. A., and J. H. Rappole. 1994. Relative homing abilities of migrants and residents in tropical rain forest of southern Veracruz, Mexico. *Bird Conservation International* 4:175–180.

Ramos O., M. A., and D. W. Warner. 1980. Analysis of North American subspecies of migrant birds wintering in Los Tuxtlas, southern Veracruz, Mexico. In *Migrant birds in the Neotropics: Ecology, behavior, distribution, and conservation,* ed. A. Keast and E. S. Morton, 173–180. Smithsonian Institution Press, Washington, D.C.

Rand, A. L. 1948. Glaciation, an isolating factor in speciation. *Evolution* 2:324–321.

Rappole, J. H. 1976. A study of evolutionary tactics in populations of solitary avian migrants. Ph.D. dissertation, University of Minnesota, Minneapolis.

————. 1983. Analysis of plumage variation in the Canada warbler. *Journal of Field Ornithology* 54:152–159.

————. 1988. Intra- and intersexual competition in migratory passerine birds during the nonbreeding season. *Proceedings of the International Ornithological Congress* 19:2308–2317.

————. 1991. Conservation priorities for migrant birds in the Neotropics. In *Conserving migratory birds,* ed. T. Salathe, 259–277. International Council for Bird Preservation, Cambridge.

Rappole, J. H., and K. Ballard. 1987. Passerine post-breeding movements in a Georgia old field community. *Wilson Bulletin* 99:475–480.

Rappole, J. H., and G. W. Blacklock. 1985. *Birds of the Texas coastal Bend.* Texas A&M University Press, College Station.

Rappole, J. H., and M. V. McDonald. 1994. Cause and effect in population declines of migratory birds. *Auk* 111:652–660.

Rappole, J. H., and E. S. Morton. 1985. Effects of habitat alteration on a tropical forest community. *Ornithological Monographs* 6:1013–1021.

Rappole, J. H., and M. A. Ramos. 1985. The current status of threatened rain forest habitat of the Tuxtla Mountains with special emphasis on endangered birds and mammals. In *Primer Simposium Internacional de Fauna Silvestre,* ed. G. Arrechea González, 397–411. Secretaria de Desarollo Urbano y Ecología, Mexico City.

————. 1994. Factors affecting migratory bird routes over the Gulf of Mexico. *Bird Conservation International* 4:131–142.

Rappole, J. H., and A. R. Tipton. 1992. The evolution of avian migration in the Neotropics. *Ornitología Neotropical* 3:45–55.

Rappole, J. H., and G. Waggerman. 1986. Calling males as an index of density for breeding White-winged Doves. *Wildlife Society Bulletin* 14:151–155.

Rappole, J. H., and D. W. Warner. 1976. Relationships between behavior, physiology, and weather in avian transients at a migration stopover site. *Oecologia* 26:193–212.

————. 1980. Ecological aspects of migrant bird behavior in Veracruz, Mexico. In *Migrant birds in the Neotropics: Ecology, behavior, distribution, and conservation,* ed. A. Keast and E. S. Morton, 353–393. Smithsonian Institution Press, Washington, D.C.

Rappole, J. H., D. W. Warner, and M. A. Ramos O. 1977. Territoriality and population structure in a small passerine community. *American Midland Naturalist* 97:110–119.

Rappole, J. H., M. A. Ramos, R. J. Oehlenschlager, D. W. Warner, and C. P. Barkan. 1979. Timing of migration and route selection in North American songbirds. In *Proceedings of the First Welder Wildlife Foundation Symposium,* ed. D. Lynn Drawe, 199–214. Welder Wildlife Foundation, Sinton, Tex.

Rappole, J. H., E. S. Morton, T. E. Lovejoy III, and J. Ruos. 1983. *Nearctic avian migrants in the Neotropics.* U.S. Fish and Wildlife Service, Washington, D.C.

————. 1993b. *Aves neárticas en los neotrópicos.* Conservation and Research Center, Front Royal, Va.

Rappole, J. H., C. E. Russell, J. R. Norwine, and T. E. Fulbright. 1985. Anthropogenic pressures and impacts on marginal, Neotropical, semiarid ecosystems: The case of south Texas. *Science for the Total Environment* 55:91–99.

Rappole, J. H., M. A. Ramos, and K. Winker. 1989. Wintering Wood Thrush mortality in southern Veracruz. *Auk* 106:402–410.

Rappole, J. H., E. S. Morton, and M. A. Ramos. 1992. Density, philopatry, and population estimates for songbird migrants wintering in Veracruz. In *Ecology and conservation of Neotropical migrant landbirds,* ed. J. M. Hagan III and D. W. Johnston, 337–344. Smithsonian Institution Press, Washington, D.C.

Rappole, J. H., W. J. McShea, and J. H. Vega-Rivera. 1993a. Estimation of species and numbers in upland avian breeding communities. *Journal of Field Ornithology* 64:55–70.

Rappole, J. H., G. V. N. Powell, and S. A. Sader. 1994. Remote sensing of tropical habitat availability for a Nearctic migrant: The Wood Thrush. In *Use of remote sensing in conservation,* ed. R. Miller, 89–103. Chapman Hall, New York.

Rappole, J. H., M. A. Ramos, K. Winker, R. J. Oehlenschlager, and D. W. Warner. n.d. Nearctic avian migrants of the Tuxtla Mountains and neighboring lowlands. In *Natural history of the Tuxtla Mountain region of southern Veracruz,* ed. R. Dirzo and R. Vogt. University of Mexico, Mexico City. In press.

Raven, P. H. 1976. The destruction of the Tropics. *Frontiers* 40:22–23.

Rea, A. 1970. Winter territoriality in a Ruby-crowned Kinglet. *Western Bird Bander* 45:4–7.

Recher, H. F., and J. T. Recher. 1969. Some aspects of the ecology of migrant shorebirds. 2. Aggression. *Wilson Bulletin* 81:140–154.

Remsen, J. V., Jr., and E. S. Hunn. 1979. First records of *Sporophila caerulescens* from Colombia: A probable long distance migrant. *Bulletin of the British Ornithological Club* 99:24–26.

Remsen, J. V., Jr., and T. A. Parker III. 1990. Seasonal distribution of the Azure Gallinule (*Porphyrula flavirostris*) with comments on vagrancy in rails and gallinules. *Wilson Bulletin* 102:380–399.

Richardson, W. J. 1976. Autumn migration over Puerto Rico and the Western Atlantic: A radar study. *Ibis* 118:309–332.

————. 1990. Timing of bird migration in relation to weather: An updated review. In *Bird migration: Physiology and ecophysiology,* ed. E. Gwinner, 78–101. Springer-Verlag, Berlin.

Ricklefs, R. E. 1972. Latitudinal variation in breeding productivity of the Rough-winged Swallow. *Auk* 89:826–936.

Ridgely, R. S. 1976. *A guide to the birds of Panama.* Princeton University Press, Princeton, N.J.

Ridgely, R. S., and G. Tudor. 1989. *The birds of South America.* Vol. 1, *The oscine passerines.* University of Texas Press, Austin.

Ridgway, R. 1901–1950. *The birds of North and Middle America*. Bulletin of the U.S. National Museum, no. 50, parts 1–11. (Parts 9–11 continued by Herbert Fried-mann.) Washington, D.C.

Ringrose, S., and W. Matheson. 1985. Consideration of effective indicators for monitoring desertification in Botswana using satellite systems. In *Proceedings of the symposium on remote sensing: Data acquisition, management, and application,* 101–111. Remote Sensing Society and CERMIA, London.

Risebrough, R. W., and D. B. Peakall. 1988. The relative importance of the several organochlorines in the decline of Peregrine Falcon populations. In *Peregrine Falcon populations,* ed. T. J. Cade, J. H. Enderson, C. G. Thelander, and C. M. White, 449–462. Peregrine Fund, Boise, Idaho.

Robbins, C. S. 1979. Effect of forest fragmentation on bird populations. In *Proceedings of the workshop on management of north central and northeastern forests for nongame birds,* ed. R. M. DeGraff and K. E. Evans, 198–212. U.S. Department of Agriculture, Forest Service, Report no. GTR NC-51. Washington, D.C.

———. 1980. Effect of forest fragmentation on breeding bird populations in the Piedmont of the mid-Atlantic region. *Atlantic Naturalist* 33:31–36.

Robbins, C. S., D. Bridge, and R. Feller. 1959. Relative abundance of adult male redstarts at an inland and a coastal locality during fall migration. *Maryland Birdlife* 15:23–25.

Robbins, C. S., D. Bystrak, and P. H. Geissler. 1986. *The breeding bird survey: Its first 15 years, 1965–1979*. U.S. Fish and Wildlife Service Publication no. 157. Washington, D.C.

Robbins, C. S., B. A. Dowell, D. K. Dawson, J. Colon, F. Espinoza, J. Rodriguez, R. Sutton, and T. Vargas. 1987. Comparison of Neotropical winter bird populations in isolated patches versus extensive forest. *Acta Oecologia Generalis* 8:285–292.

Robbins, C. S., D. K. Dawson, and B. A. Dowell. 1989a. Habitat area requirements of breeding forest birds of the Middle Atlantic states. *Wildlife Monographs,* no. 103.

Robbins, C. S., J. R. Sauer, R. Greenberg, and S. Droege. 1989b. Population declines in North American birds that migrate to the Neotropics. *Proceedings of the National Academy of Sciences* (USA) 86:7658–7662.

Robbins, C. S., J. W. Fitzpatrick, and P. B. Hamel. 1992a. A warbler in trouble: *Dendroica cerulea*. In *Ecology and conservation of Neotropical migrant landbirds,* ed. J. M. Hagan III and D. W. Johnston, 549–562. Smithsonian Institution Press, Washington, D.C.

Robbins, C. S., with collaborators. 1992b. Comparison of Neotropical migrant landbird populations wintering in tropical forest, isolated forest fragments, and agricultural habitats. In *Ecology and conservation of Neotropical migrant landbirds,* ed. J. M. Hagan III and D. W. Johnston, 207–220. Smithsonian Institution Press, Washington, D.C.

Robertson, W. B., Jr. 1962. Observations on the birds of St. John, Virgin Islands. *Auk* 79:44–76.

Robinson, S. K. 1992. Population dynamics of breeding Neotropical migrants in a fragmented Illinois landscape. In *Ecology and conservation of Neotropical migrant landbirds,* ed. J. M. Hagan III and D. W. Johnston, 408–418. Smithsonian Institution Press, Washington, D.C.

Robinson, S. K., J. Terborgh, and J. W. Fitzpatrick. 1988. Habitat selection and relative abundance of migrants in southeastern Peru. *Proceedings of the International Ornithological Congress* 19:2298–2307.

Roche, M. A. 1977. Hydrodynamics and evaluation of pollution risk in a tidal estuary (French Guiana). *Cahiers ORSTOM, Série Hydrologie* 14:345–382.

Rodriguez, C. 1972. Mechanisms of action for protecting tropical estuaries [in Spanish]. In *Simposio internacional sobre la protección del medio ambiente y los recursos naturales,* 395–406.

Rohwer, S. 1975. The social significance of avian winter plumage variability. *Evolution* 29:593–610.

Roubal, G., and R. M. Atlas. 1980. Biodegradation of crude oil mousse in the Gulf of Mexico from the Ixtoc well blowout [abstract]. Annual Meeting of the American Society of Microbiologists.

Sader, S. A., and A. T. Joyce. 1988. Deforestation rates and trends in Costa Rica, 1940 to 1983. *Biotropica* 20:11–19.

Sader, S. A., G. V. N. Powell, and J. H. Rappole. 1991. Migratory bird habitat monitoring through remote sensing. *International Journal of Remote Sensing* 12:363–372.

Salas, G. de las, and H. Folster. 1976. Bioelement loss on clearing a tropical rain forest. *Turrialba* 26:179–186.

Salomonsen, F. 1955. Evolution and bird migration. *Proceedings of the International Ornithological Congress* 11:337–339.

———. 1968. The moult migration. *Wildfowl* 19:5–24.

Santos, P. L., and A. de G. Dos Pedrini. 1978. Preliminary results on the accumulation and loss of radionucleides in marine benthic algae [in Portuguese]. *Rodriguesia* (Brazil) 29:263–268.

Sauer, J. R., and S. Droege. 1992. Geographic patterns in population trends of Neotropical migrants in North America. In *Ecology and conservation of Neotropical migrant landbirds,* ed. J. M. Hagan III and D. W. Johnston, 26–42. Smithsonian Institution Press, Washington, D.C.

Sauer, J. R., B. G. Peterjohn, and W. A. Link. 1994. Observer differences in the North American Breeding Bird Survey. *Auk* 111:50–62.

Saunders, G. B. 1952. Waterfowl wintering grounds of Mexico. *Transactions of the North American Wildlife and Natural Resources Conference* 17:89–100.

Schlatter, R. P. 1976. Birds observed in a sector of Lake Pinihue, Valdivia Province, Chile, with data on their ecology. *Boletín de la Sociedad de Biológica Conservación* 50:133–144.

Scholander, S. I. 1955. Land-birds over the western North Atlantic. *Auk* 72: 225–240.

Schwartz, P. 1963. Orientation experiments with Northern Waterthrushes wintering in Venezuela. *Proceedings of the International Ornithological Congress* 13:481–484.

———. 1964. The Northern Waterthrush in Venezuela. *Living Bird* 3:169–184.

————. 1980. Some considerations on migratory birds. In *Migrant birds in the Neotropics: Ecology, behavior, distribution, and conservation,* ed. A. Keast and E. S. Morton, 31–34. Smithsonian Institution Press, Washington, D.C.

Senner, S. E. 1993. Frequent flyers—destination Neotropics. *World Birdwatch* 15(3): 6–8.

Serrao, J. 1985. Decline of forest songbirds. *Records of New Jersey Birds* 11:5–9.

Serrentino, P. 1992. Northern Harrier (*Circus cyaneus*). In *Migratory nongame birds of management concern in the northeast,* ed. K. J. Schneider and D. M. Pence, 89–118. U.S. Fish and Wildlife Service, Region 5, Newton Corner, Mass.

Sharp, B. 1985. Avifaunal changes in central Oregon since 1899. *Western Birds* 16:63–70.

Sherry, T. W., and R. T. Holmes. 1992. Population fluctuations in a long-distance Neotropical migrant: Demographic evidence for the importance of breeding season events in the American Redstart. In *Ecology and conservation of Neotropical migrant landbirds,* ed. J. M. Hagan III and D. W. Johnston, 431–442. Smithsonian Institution Press, Washington, D.C.

Sick, H. 1968. Bird migration in continental South America [in German]. *Vogelwarte* 24:217–243.

————. 1984. *Brazilian ornithology: An introduction* [in Portuguese]. Editoria Universidade de Brasília, Brazil.

Skutch, A. F. 1944. Life history of the Quetzal. *Condor* 46:213–235.

————. 1950. The nesting season of Central American birds in relation to climate and food supply. *Ibis* 92:185–222.

————. 1976. *Parent birds and their young.* University of Texas Press, Austin.

————. 1980. Arils as food of tropical American birds. *Condor* 82:31–42.

Slatkin, M. 1984. Ecological causes of sexual dimorphism. *Evolution* 38:622–630.

Slud, P. 1960. The Birds of Finca "La Selva" Costa Rica: A tropical wet forest locality. *Bulletin of the American Museum of Natural History* 121:49–148.

————. 1964. The birds of Costa Rica. *Bulletin of the American Museum of Natural History* 128:1–430.

Smith, N. G. 1975. Spshing noise: Biological significance of its attraction and non-attraction by birds. *Proceedings of the National Academy of Sciences* (USA) 72: 1411–1414.

————. 1980. Hawk and vulture migrations in the Neotropics. In *Migrant birds in the Neotropics: Ecology, behavior, distribution, and conservation,* ed. A. Keast and E. S. Morton, 51–65. Smithsonian Institution Press, Washington, D.C.

Snow, D. W., and B. K. Snow. 1960. Northern Waterthrush returning to same winter quarters in successive winters. *Auk* 77:351–352.

————. 1964. Breeding seasons and annual cycles of Trinidad land birds. *Zoologica* 49:1–30.

Snyder, D. E. 1966. *The birds of Guyana.* Peabody Museum of Natural History, Salem, Mass.

Snyder, N. F. R., and J. W. Wiley. 1976. Sexual size dimorphism in hawks and owls of North America. *Ornithological Monographs,* no. 20.

Sommer, A. 1976. Attempt at an assessment of the world's tropical moist forests. *Unasylva* 28:5–24.

Sorola, S. H. 1984. *Investigation of Mearn's Quail distribution*. Texas Parks and Wildlife Department, Federal Aid Performance Report, Project W-108-R-7. Austin, Tex.

Sosa Ferreyro, R. A. 1976. Without forests the future of Mexico would be characterized by erosion and deserts [in Spanish]. *Mexican Forestry* 50:9–11.

Staicer, C. A. 1992. Social behavior of the Northern Parula, Cape May Warbler, and Prairie Warbler wintering in second-growth forest in southwestern Puerto Rico. In *Ecology and conservation of Neotropical migrant landbirds,* ed. J. M. Hagan III and D. W. Johnston, 308–320. Smithsonian Institution Press, Washington, D.C.

Steidl, R. J., C. R. Griffin, L. J. Niles, and K. E. Clark. 1991. Reproductive success and eggshell thinning of a reestablished Peregrine Falcon population. *Journal of Wildlife Management* 55:294–299.

Stepney, P. H. R. 1975. First recorded breeding of the Great-tailed Grackle in Colorado. *Condor* 77:208–210.

Stevenson, H. M. 1957. On the relative magnitude of the trans-Gulf and circum-Gulf spring migration. *Wilson Bulletin* 69:39–77.

Stewart, P. A. 1987. Decline in numbers of wood warblers in spring and autumn migrations through Ohio. *North American Bird Bander* 12:58–60.

Stewart, R. M., L. R. Mewaldt, and S. Kaiser. 1974. Age ratios of coastal and inland fall migrant passerines in central California. *Bird-Banding* 45:46–57.

Stewart, R. M., R. P. Henderson, and K. Darling. 1977. Breeding ecology of Wilson's Warbler in the High Sierra Nevada, California. *Living Bird* 16:83–102.

Stiles, F. G. 1976. *Checklist of the birds of La Selva and vicinity.* University of Costa Rica, Heredia. Mimeograph.

———. 1980. Evolutionary implications of habitat relations between permanent resident and winter resident land birds in Costa Rica. In *Migrant birds in the Neotropics: Ecology, behavior, distribution, and conservation,* ed. A. Keast and E. S. Morton, 421–435. Smithsonian Institution Press, Washington, D.C.

———. 1983. Birds. In *Costa Rican natural history,* ed. D. H. Janzen, 502–530. University of Chicago Press, Chicago.

———. 1985a. On the roles of birds in the dynamics of Neotropical forests. In *Conservation of tropical forest birds,* ed. A. W. Diamond and T. E. Lovejoy III, 49–59. International Council for Bird Preservation, Cambridge.

———. 1985b. Seasonal and altitudinal changes in the avifauna on the Atlantic slope of Costa Rica [in Spanish]. In *First Symposium for Neotropical Ornithology* [in Spanish], ed. F. G. Stiles and P. G. Aguilar, 95–104. Asociación Peruana Conservación Natural, Lima, Peru.

———. 1988. Notes on the distribution and status of certain birds in Costa Rica. *Condor* 90:931–933.

Stiles, F. G., and S. M. Smith. 1977. New information on Costa Rican waterbirds. *Condor* 79:91–97.

Stone, C. P. 1966. *A literature review on Mourning Dove song as related to the coo-count census.* Colorado Department of Game, Fish, and Parks, Special Report no. 11.

Stutchbury, B. J. 1994. Competition for winter territories in a Neotropical migrant: The role of age, sex, and color. *Auk* 111:63–69.

Sutton, G. M. 1944. The kites of the genus *Ictinia. Wilson Bulletin* 56:3–8.

Swanson, D. A. 1989. Breeding biology of the White-winged Dove (*Zenaida asiatica*) in south Texas. Master's thesis, Texas A&I University, Kingsville.

Tate, G. R. 1992. Short-eared Owl (*Asio flammeus*). In *Migratory nongame birds of management concern in the northeast,* ed. K. J. Schneider and D. M. Pence, 171–190. U.S. Fish and Wildlife Service, Region 5, Newton Corner, Mass.

Taverner, P. A. 1904. A discussion of the origin of migration. *Auk* 21:322–333.

Temeles, E. J. 1985. Sexual size dimorphism of bird-eating hawks: The effect of prey vulnerability. *American Naturalist* 125:485–499.

Temple, S. A., and J. R. Cary. 1988. Modelling dynamics of habitat-interior bird populations in fragmented landscapes. *Conservation Biology* 2:340–347.

Temple, S. A., and B. L. Temple. 1976. Avian population trends in central New York State, 1935–1973. *Bird-Banding* 47:238–257.

Terborgh, J. W. 1989. *Where have all the birds gone?* Princeton University Press, Princeton, N.J.

Terborgh, J. W., and J. S. Weske. 1969. Colonization of secondary habitats by Peruvian birds. *Ecology* 50:765–782.

Terrill, S. B. 1988. The relative importance of ecological factors in bird migration. *Proceedings of the International Ornithological Congress* 19: 2180–2190.

———. 1990. Food availability, migratory behavior, and population dynamics of terrestrial birds during the nonreproductive season. *Studies in Avian Biology* 13: 438–443.

Thiollay, J. M. 1970a. Avian populations of a scrub-savannah (Lamto, Ivory Coast) [in French]. Ph.D. dissertation, Abidjan University, Abidjan, Ivory Coast.

———. 1970b. Ecological research in the savannah of Lamto (Ivory Coast): Avian populations [in French]. *Terre et Vie* 24:108–144.

———. 1977. Autumn migration along the eastern coast of Mexico. *Alauda* 45: 344–346.

———. 1979. Importance of an axis of migration along the east coast of Mexico [in French]. *Alauda* 47:235–245.

———. 1988. Comparative foraging success of insectivorous birds in tropical and temperate forests: Ecological implications. *Oikos* 53:17–30.

Thompson, C. F., and V. Nolan, Jr. 1973. Population biology of the Yellow-breasted Chat (*Icteria virens* L.) in southern Indiana. *Ecological Monographs* 43:145–171.

Thurber, W. A., and A. Villeda C. 1972. Some banding returns in El Salvador. *Bird-Banding* 43:285.

———. 1976. Band returns in El Salvador, 1973–74 and 1974–75 seasons. *Bird-Banding* 47:277–278.

Titus, K. 1990. Trends in counts of Scissor-tailed Flycatchers based on a nonparametric rank-trend analysis. In *Survey designs and statistical methods for the estimation of avian population trends,* ed. J. R. Sauer and S. Droege, 164–166. U.S. Fish and Wildlife Service, Biological Reports, vol. 90, no. 1. Washington, D.C.

Titus, K., M. R. Fuller, and D. Jacobs. 1990. Detecting trends in hawk migration count data. In *Survey designs and statistical methods for the estimation of avian population trends,* ed. J. R. Sauer and S. Droege, 105–113. U.S. Fish and Wildlife Service, Biological Reports, vol. 90, no. 1. Washington, D.C.

Tosi, J. A. 1963. The natural resources of Latin America: Possibilities of integration [in Spanish]. In *Asemblea Nacional de Conservación de los Recursos Naturales Renovables, 1ª, Caracas, 1962: Ponencias y resoluciones,* 339–344. Ministerio de Agricultura y Cría, Caracas, Venezuela.

Tramer, E. J. 1974. Proportions of wintering North American birds in disturbed and undisturbed dry tropical habitats. *Condor* 76:460–464.

Tramer, E. J., and T. R. Kemp. 1980. Foraging ecology of migrants and resident warblers and vireos in the highlands of Costa Rica. In *Migrant birds in the Neotropics: Ecology, behavior, distribution, and conservation,* ed. A. Keast and E. S. Morton, 285–296. Smithsonian Institution Press, Washington, D.C.

UNESCO. 1978. *Tropical forest ecosystems: A state-of-knowledge report.* Natural Resources Research 14. UNESCO, Paris.

U.S. Fish and Wildlife Service. 1987. *Migratory nongame birds of management concern in the United States: The 1987 list.* U.S. Department of the Interior, Office of Migratory Bird Management, Washington, D.C.

U.S. National Oceanic and Atmospheric Administration. 1979. *Virgin Islands coastal management program: Coastal zone management plan for Virgin Islands.* U.S. National Oceanic and Atmospheric Administration, Department of Commerce, Washington, D.C.

U.S. State Department. 1980. *The world's tropical forests: A policy, strategy, and program for the United States.* International Organizations and Conferences Series 145. U.S. Publication no. 9117.

Van Devender, T. R. 1990. Late Quaternary vegetation and climate of the Sonoran Desert, United States and Mexico. In *Packrat middens: The last 40,000 years of biotic change,* ed. J. L. Betancourt, T. R. Van Devender, and P. S. Martin, 104–133. University of Arizona Press, Tucson.

Van Horne, B. 1983. Density as a misleading indicator of habitat quality. *Journal of Wildlife Management* 47:893–901.

van Rossem, A. J. 1936. Description of a race of *Myiarchus cinerascens* from El Salvador. *Transactions of the San Diego Society of Natural History* 8:115–118.

Van Tyne, J. 1932. Winter returns of the Indigo Bunting in Guatemala. *Bird-Banding* 3:110.

Vaurie, C. 1959, 1965. *The birds of the Palaearctic avifauna.* 2 vols. Witherby, London.

Vega, J. H., and J. H. Rappole. 1994. Composition and phenology of an avian community in the Rio Grande Plain of Texas. *Wilson Bulletin* 106:366–380.

Vermeer, K., R. W. Risebrough, A. L. Spaans, and L. M. Reynolds. 1974. Pesticide effects on fishes and birds in rice fields of Surinam, South America. *Environmental Pollution* 7:217–236.

Verner, J., and M. F. Willson. 1969. Mating systems, sexual dimorphism, and the role of male North American passerine birds in the nesting cycle. *Ornithological Monographs,* no. 9.

Via, J., and D. C. Duffy. 1992. Northern Harrier (*Circus cyaneus*). In *Migratory nongame birds of management concern in the northeast,* ed. K. J. Schneider and D. M. Pence, 89–118. U.S. Fish and Wildlife Service, Region 5, Newton Corner, Mass.

Vickery, P. D., M. L. Hunter, Jr., and J. V. Wells. 1992. Is density an indicator of breeding success? *Auk* 109:706–710.

Villard, M., P. R. Martin, and C. G. Drummond. 1993. Habitat fragmentation and pairing success in the Ovenbird (*Seiurus aurocapillus*). *Auk* 110:759–768.

Vogt, W. 1970. The avifauna in a changing ecosystem. In *A symposium of the Smithsonian Institution on the avifauna of northern Latin America,* ed. H. K. Buechner and J. H. Buechner, 8–16. Smithsonian Contributions to Zoology, no. 26. Washington, D.C.

von Haartman, L. 1971. Population dynamics. In *Avian biology,* vol. 1, ed. D. S. Farner and J. R. King, 392–449. Academic Press, New York.

Voous, K. H. 1955. *The birds of the Netherlands Antilles.* Natuurwet, Werkgroep Nederlandse Antillen, Curaçao, Netherlands Antilles.

Wagner, H. O. 1959. Composition of bird flocks in Mexico: Especially the behavior of northern migrants [in German]. *Zeitschrift für Tierpsychologie* 15:178–190.

Waide, R. B. 1980. Resource partitioning between migrant and resident birds: The use of irregular resources. In *Migrant birds in the Neotropics: Ecology, behavior, distribution, and conservation,* ed. A. Keast and E. S. Morton, 337–352. Smithsonian Institution Press, Washington, D.C.

Walcott, C. F. 1974. Changes in bird life in Cambridge, Massachusetts from 1860 to 1964. *Auk* 91:151–160.

Wallace, A. R. 1874. Letter on migration of birds. *Nature* 10:459.

Walter, H. E. 1908. Theories of bird migration. *School of Science and Mathematics* 8:259–268, 359–366.

Ward, P., and A. Zahavi. 1973. The importance of certain assemblages of birds as "information centres" for food finding. *Ibis* 115:517–534.

Warren, R. J. 1991. Ecological justification for controlling deer populations in eastern national parks. *Transactions of the North American Wildlife and Natural Resources Conference* 56:56–66.

Webb, B. E. 1976. A Groove-billed Ani in northeastern Colorado. *Western Birds* 7:153–154.

Weber, W. C., and J. A. Jackson. 1977. First nesting record of Gray Kingbirds in Mississippi. *Mississippi Kite* 7:10–12.

Webster, J. D. 1959. A revision of the Botteri Sparrow. *Condor* 61:136–146.

Wells, M., and K. Brandon. 1992. *People and parks: Linking protected area management with local communities.* World Bank, Washington, D.C.

Wetmore, A. 1926. *The migrations of birds.* Harvard University Press, Cambridge.

———. 1943. The birds of southern Veracruz, Mexico. *Proceedings of the U.S. National Museum* 93:215–340.

Wetmore, A., and B. H. Swales. 1931. *The birds of Haiti and the Dominican Republic.* Bulletin of the U.S. National Museum, no. 155. Washington, D.C.

Weyl, R. 1972. Forest devastation, soil erosion, and water supply in Colombia. *National Museum* 102:292–300.

Wheelwright, N. T. 1983. Fruits and the ecology of Resplendent Quetzals. *Auk* 100: 286–301.

Wheelwright, N. T., W. A. Haber, K. G. Murray, and C. Guindon. 1984. Tropical fruit-eating birds and their food plants: A survey of a Costa Rican lower montane forest. *Biotropica* 16:173–192.

Whitcomb, R. F. 1977. Island biogeography and "habitat islands" of eastern forest. *American Birds* 31:3–5.

Wiedenfeld, D. A. 1992. Foraging in temperate- and tropical-breeding and wintering male Yellow Warblers. In *Ecology and conservation of Neotropical migrant landbirds,* ed. J. M. Hagan III and D. W. Johnston, 321–328. Smithsonian Institution Press, Washington, D.C.

Wiens, J. A. 1977. On competition and variable environments. *American Scientist* 65:590–597.

Wierenga, H. 1978. Maryland's first Fork-tailed Flycatcher. *Maryland Birdlife* 34: 171–172.

Wilcove, D. S. 1983. Population changes in the Neotropical migrants of the Great Smoky Mountains: 1947–1982. Unpublished report to the World Wildlife Fund, USA.

———. 1985. Nest predation in forest tracts and the decline of migratory songbirds. *Ecology* 66:1211–1214.

Wildash, P. 1968. *The birds of South Vietnam.* C. E. Tuttle, Rutland, Vt.

Willcox, D. R. C., and B. Willcox. 1978. Observations of birds in Tripolitania, Libya. *Ibis* 120:329–332.

Williams, G. G. 1945. Do birds cross the Gulf of Mexico in spring? *Auk* 62:98–111.

———. 1947. Lowery on trans-Gulf migration. *Auk* 64:217–238.

———. 1950. The nature and causes of the coastal hiatus. *Wilson Bulletin* 62: 175–182.

———. 1951. Letter to the editor. *Wilson Bulletin* 63:52–54.

———. 1958. *Evolutionary aspects of bird migration.* Lida Scott Brown Lectures in Ornithology, University of California, Los Angeles, 53–85.

Williams, T. C., and J. M. Williams. 1978. Orientation of trans-Atlantic migrants. In *Symposium on animal migration navigation and homing,* ed. K. Schmidt-Koenig and W. T. Keeton, 239–251. Springer-Verlag, New York.

Williamson, P. 1971. Feeding ecology of the Red-eyed Vireo (*Vireo olivaceus*) and associated foliage gleaning birds. *Ecological Monographs* 41:129–152.

Willis, E. O. 1966. The role of migrant birds at swarms of army ants. *Living Bird* 5:187–231.

———. 1973. Local distribution of mixed flocks in Puerto Rico. *Wilson Bulletin* 85:75–77.

———. 1974. Populations and local extinctions of birds on Barro Colorado Island, Panama. *Ecological Monographs* 44:153–169.

———. 1980. Ecological roles of migratory and resident birds on Barro Colorado Island, Panama. In *Migrant birds in the Neotropics: Ecology, behavior, distribution, and conservation,* ed. A. Keast and E. S. Morton, 205–225. Smithsonian Institution Press, Washington, D.C.

———. 1988. Land-bird migration in São Paulo, southeastern Brazil. *Proceedings of the International Ornithological Congress* 19:754–764.

Wiltschko, W., and K. P. Able. 1988. Symposium 33: Migratory orientation. *Proceedings of the International Ornithological Congress* 19:1918–1962.

Wingate, D. B. 1973. *A checklist and guide to the birds of Bermuda.* Island Press, Hamilton, Bermuda.

Winker, K. 1989. Ecology of the Wood Thrush in southern Veracruz. Master's thesis, University of Minnesota, Minneapolis.

Winker, K., and J. H. Rappole. 1988. Taxonomic relationships between the genera *Hylocichla* and *Catharus. Auk* 105:392–394.

———. 1992. Timing of migration in the Yellow-bellied Flycatcher in south Texas. *Condor* 94:525–529.

Winker, K., J. H. Rappole, and M. A. Ramos. 1990a. Population dynamics of the Wood Thrush (*Hylocichla mustelina*) on its wintering grounds in southern Veracruz, Mexico. *Condor* 92:444–460.

———. 1990b. Within-forest preferences of Wood Thrushes wintering in the rainforest of southern Veracruz. *Wilson Bulletin* 102:715–720.

———. 1995. The use of movement data as an assay of habitat quality. *Oecologia* 101:211–216.

Winker, K., R. J. Oehlenschlager, M. A. Ramos, R. M. Zink, J. H. Rappole, and D. W. Warner. 1992a. Bird distribution and abundance records for the Sierra de los Tuxtlas, Veracruz, Mexico. *Wilson Bulletin* 104:699–718.

Winker, K., D. W. Warner, and A. R. Weisbrod. 1992b. Daily mass gains among woodland migrants at an inland stopover site. *Auk* 109:853–862.

———. 1992c. Migration of woodland birds at a fragmented inland stopover site. *Wilson Bulletin* 104:580–598.

———. 1992d. The Northern Waterthrush and Swainson's Thrush as transients at a temperate inland stopover site. In *Ecology and conservation of Neotropical migrant landbirds,* ed. J. M. Hagan III and D. W. Johnston, 384–402. Smithsonian Institution Press, Washington, D.C.

Winker, K., G. A. Voelker, and J. T. Klicka. 1994. A morphometric examination of sexual dimorphism in the *Hylophilus, Xenops,* and an *Automolus* from southern Veracruz, Mexico. *Journal of Field Ornithology* 65:307–323.

Witham, J. W., and M. L. Hunter, Jr. 1992. Population trends of Neotropical migrant landbirds in northern coastal New England. In *Ecology and conservation of*

Neotropical migrant landbirds, ed. J. M. Hagan III and D. W. Johnston, 85–95. Smithsonian Institution Press, Washington, D.C.

Wittenberger, J. F. 1980. Vegetation structure, food supply, and polygyny in Bobolinks (*Dolichonyx oryzivorus*). *Ecology* 61:140–150.

Witzeman, J. 1979. Plain-capped Starthroats in the United States. *Continental Birdlife* 1:1–3.

Wolf, D. E. 1978. First record of an Aztec Thrush in the United States. *American Birds* 32:156–157.

Wolfson, A. 1948. Bird migration and the concept of continental drift. *Science* 108: 23–30.

Woods, C. A. 1975. Banding and recapture of wintering Warblers in Haiti. *Bird-Banding* 46:344–346.

World Resources Institute. 1992. *World resources: 1992–1993.* Oxford University Press, New York.

Wunderle, J. M., Jr. 1992. Sexual habitat segregation in wintering Black-throated Blue Warblers in Puerto Rico. In *Ecology and conservation of Neotropical migrant landbirds,* ed. J. M. Hagan III and D. W. Johnston, 299–307. Smithsonian Institution Press, Washington, D.C.

Zahavi, A. 1971. The social behavior of the white wagtail *Motacilla alba alba* wintering in Israel. *Ibis* 113:203–211.

Zimmer, J. T. 1931–1945. *Studies of Peruvian birds: Revision of Picidae, Rhinocryptidae, Tyrannidae, Vireonidae, Coerebidae, Thraupidae, Formicariidae, Sylviidae.* American Museum Novitates, unnumbered index, 31 December 1945.

———. 1938. Notes on migrations of South American birds. *Auk* 55:405–410.

———. 1947–1955. *Studies of Peruvian birds: Revision of Trogonidae, Parulidae, Trochilidae, Apodidae, Motacillidae, Thraupidae.* American Museum Novitates, unnumbered index, 1957.

Zimmerman, D. A. 1978. Eared Trogon—immigrant or visitor? *American Birds* 32:135–139.

Zuvekas, C. 1978. *Agriculture development in Haiti: An assessment of sector problems, policies, and prospects under conditions of severe soil erosion.* Agency for International Development, Washington, D.C.

INDEX OF SCIENTIFIC NAMES

English common names used in the book are given in parentheses after the scientific name. Page numbers for tables and illustrations are in italics; page numbers for Appendixes 1–3 are in bold. English and Spanish common names, with corresponding scientific names and Spanish or English counterparts, are listed in Appendixes 4 and 5, respectively.